selected topics in biology

FOOD CHAINS TO BIOTECHNOLOGY

J.B. Land B.Sc. Ph.D.
Marlborough College, Wiltshire

R.B. Land B.Sc. Ph.D.
ARC Animal Breeding Research Organisation, Edinburgh

Nelson

Thomas Nelson and Sons Ltd
Nelson House Mayfield Road Walton-on-Thames Surrey KT12 5PL

51 York Place Edinburgh EH1 3JD

Thomas Nelson (Hong Kong) Ltd
Toppan Building 10/F 22A Westlands Road
Quarry Bay Hong Kong

Distributed in Australia by

Thomas Nelson Australia
480 La Trobe Street
Melbourne Victoria 3000
and in Sydney, Brisbane, Adelaide and Perth

First published by Thomas Nelson and Sons Ltd 1983

ISBN 0-17-448088-1
NPN 9 8 7 6 5

Illustrated by Illustrated Arts Ltd, Sutton, Surrey
Drawings on pages 193 and 194 by Claire Halliday
Printed and bound in Hong Kong

General Editor's Preface

The books in this series are written specifically for Advanced-level students who wish to pursue certain areas of biology in depth. We hope the books will be particularly useful to those who are taking special papers or entrance examinations to Oxford or Cambridge.

The writers of books of this kind face a dilemma. On the one hand they want to provide plenty of material to challenge the keen sixth former. On the other hand they do not want to smother the poor student with a plethora of detail which he or she will shortly meet again at university. The aim of this series is to broaden the student's view of biology without trespassing on university territory. The emphasis is on providing a wide range of interesting examples and studies which we hope will stimulate thought and enquiry. At the same time we hope that the books will prove enjoyable as well as informative.

In preparing the series we have been highly selective in our choice of topics. We have confined ourselves to those topics which are difficult to cover adequately in a basic textbook and for which we consider there is a need for appropriate books at the sixth form level.

M. B. V. Roberts

Authors' preface

Biology, and so the biologist, offers a range of solutions to many of the present and future needs of humankind; this book is written to bring some of these possibilities to the attention of the Advanced-level student.

Another aim is to present to the student a sequence of biological principles which are firmly anchored in the Advanced-level course, so this book will both *complement* and *run parallel* to a standard Advanced-level text. We explore the principles, using examples of plants and animals as organisms producing materials for humans. These illustrate the particular principle in the context of contemporary trends in production, and at a level suitable for the Advanced-level student. We do not intend this to be a handbook for production.

Two examples will make clear our intention.

Consider the flow of energy through an ecosystem. Autotrophic organisms partition photosynthate with varying efficiency and proportion into starch and cellulose. Starch is readily used by humans and most animals, but it could equally well be used as an energy base to make alcohol for cars. Cellulose is used with low efficiency and then only by those animals, such as ruminants, which have micro-organisms in their gut. This cellulosic material represents a vast untapped source of energy. Pretreatment before it is fed to ruminants improves its digestibility, and micro-organisms could upgrade it to produce a nutritious animal feed. Manipulation of the genetic make-up of the micro-organism growing in the gut could improve the ruminant's use of cellulose. How can we best improve the flow of energy for the benefit of humans?

The second example is a development of the consideration of the role of insulin, from slaughtered animals, in alleviating the symptoms of diabetes mellitus, a principle well rooted in Advanced-level courses. The human immunological recognition systems distinguish between human, pig and cow insulins and so a need for human insulin is acknowledged. This enables us to illustrate the principles of genetic engineering and microbiological production of compounds such as insulin which are of value to humans; we can also reflect upon the economic principles of developing the production facilities.

We feel that the combinations of a schoolmaster and a practising research worker, and of a plant scientist with an animal scientist, are complementary and we hope that the book reflects this. We hope that it is readable, yet gives a flavour of the biological problems (and some of the answers) facing humankind over the next 25 years, that it is firmly rooted in basic biological principles and yet is right up to date, and that it might be profitably used early in an Advanced-level course and yet remain relevant throughout the course.

We hope that reading the book will help to clarify the principles of biology, but moreover that it will stimulate the student to respond to the challenge of biology in the service of society.

J.B. Land
R.B. Land

Acknowledgements

We cannot thank too warmly those many scientists, often friends, colleagues or teachers, whose papers and books we have read, whose lectures and seminars we have listened to, who responded to our teaching, research and publications, and who talked biological shop; over the years these scientific interactions have formulated our own ideas and research.

Dr Ratledge, of the University of Hull, was particularly helpful initially in introducing specialist biotechnology literature to us. More recently, Dr R. Tilbury, of Tate and Lyle Ltd, British Petroleum Ltd, and Imperial Chemical Industries Ltd have provided research and commercial literature. Agricultural Research Council Information Officers responded very helpfully and willingly to general and specific queries about research in their Institutes.

We are most grateful to Ellen Firth and Carol Manos who typed the original manuscript with such care. But for Michael Roberts, the project would not have started; his ideas and his example as a teacher and writer, along with his continued encouragement, have been a major reason for expressing our ideas in this form and as an A-level book.

Elizabeth Johnston and Andrew Nash of Nelson willingly and skilfully helped the original manuscript into the world as a book.

This book is a family production: we hope that this will offer some consolation to our families for the disruption of our holidays.

Contents

1 Introduction – the problems

Wild populations of plants and animals evolve under the forces of **natural selection**, as proposed by Charles Darwin in *On the Origin of Species* (published in 1859). In a population, it is the fittest individuals – those best suited to their environment – which survive; and it is the individuals which survive that have offspring, and so pass on their genes to the next generation.

The same is not true of cultivated plants and animals, because humans protect certain individuals from the forces of nature which might otherwise destroy them. Selection is no longer natural, it is by deliberate choice of humans; and the criterion is no longer fitness in the natural environment, but usefulness to humans. (Of course, it is possible to consider humans as just one more agent in the natural environment!)

This book is about the way in which humans use their biological knowledge to cultivate plants and animals which they can use, most often for food. It is important to be clear at the outset about a fundamental issue. When looking at a plant or animal, what is visible – characteristics such as root length, seed size, milk yield and so on – is the result, the *expression*, of the genes in that organism. The organism has a set of genes which constitute its **genotype**, and these, in conjunction with the effects of the environment, give rise to its outward appearance, the **phenotype**. Plant and animal breeders are always concerned with influencing the phenotype, yet they must do so by affecting the genotype – to alter what they can see, they must alter what they cannot see. And in fact *we* can rarely alter genes; we rely on nature to provide us with alternatives, and we simply select the ones which seem preferable.

Our interference with natural selection is of two sorts: we can influence the environment in which organisms grow, and we can influence the way in which they breed, and thus affect the next generation. Influence may be as minor as the introduction into a flock of mountain sheep of one particular ram, or as major, elaborate and extensive as exercising complete control over the organism's entire life, as with poultry production units. The object is always the same: to improve the characteristics of the organism to benefit ourselves.

Choice of characteristic

For a given organism, how do we decide which **characteristics** to improve, supposing that we can improve them? Do we try for increased height, or weight, or speed of growth, or what? In practice, the decision is not always the obvious one, and is often dictated by economics: how much society is prepared to pay.

In general terms, then, we assess the biological characteristics of the organism, and their value to the consumer. Three points emerge. Firstly, the characteristics then chosen are often the subtler ones – for example, it may be less important to grow large raspberries than to grow raspberries which can be harvested mechanically. Secondly, improvements in one respect may produce disadvantages in another – for instance, a ewe may be persuaded to have more lambs, but these lambs may grow more slowly. Thirdly, economic values may be limited to one part of the world only – pigeon fancying, for example, may be of great interest in the north of England but of none whatsoever in, say, Bangladesh.

As explained above, the phenotypic characteristic is controlled by unseen genes, and it is not yet easy to tamper with genes directly. Tools are becoming more sensitive, as we shall see when we look at so-called genetic engineering, but we still rely to a great extent on natural **variation:** the fact that individuals within a population differ from each other. If there is no variation, or if the variation shows no superiority of one over another, there can be no selection, and so no controlled genetic improvement.

Efficiency in improvement

One aspect of improvement is that of increasing efficiency – efficiency of energy use in food, for example. The criteria are by no means clear or simple, as is demonstrated by the food surpluses which occur in the European Economic Community (EEC), such as butter 'mountains' and wine 'lakes'. From purely energetic considerations, it is more efficient for humans to consume plants than animals (as we shall see later), yet the removal of the intermediate link – the animals – would produce dramatic changes. The lower Alps, for example, would lose much of their charm if devoid of grazing cattle! Efficiency is subject to many political, cultural and economic pressures, and these may sometimes conflict.

Incentives for improvement

Today's alteration of plants and animals for the benefit of humans is merely an extension of the gradual process of domestication which has been taking place for about 10 000 years. And even this span of time accounts for only 0.05 per cent of time on earth.

Methods of improvement have become more careful and specific over the last few hundred years, and the process is fast developing. The incentives have always included the desire to improve living standards, but there is now the increasingly pressing need to maintain nutritional standards in the face of ever more people. World population and the

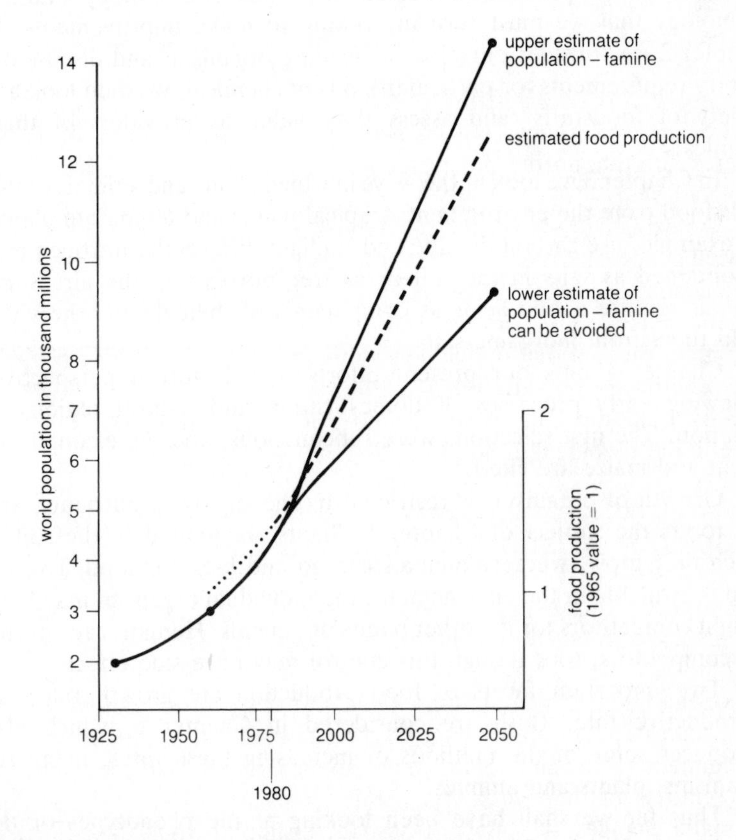

Figure 1.1 World population and food production.

demand for food are rising and a general famine condition may exist next century (Figure 1.1).

Improvements are also of political importance. The EEC, for example, is largely self-sufficient in foodstuffs: it produces a total of 105 million tonnes of grain, making it 90 per cent self-sufficient, and about 75 kg of meat per person, about 98 per cent self-sufficient. The world as a whole produces about 30 kg of meat per head. To match the predicted increase in world population, meat production would need to rise by one-half in the next 25 years, simply to stay at the same amount of meat per person; and it would need to rise vastly more to reach current EEC levels.

Contents of this book

Plant and animal production is essentially a matter of biology, and it is to biology that we must turn in seeking to make improvements. In Chapter 2 we consider ourselves as feeding organisms and discuss our dietary requirements for particular kinds of chemical; we then look at a variety of foodstuffs, and assess their value as providers of these chemicals.

In Chapter 3 we look at the ways in which plants and animals obtain their food from the environment: tropical plants and temperate plants, for example, use carbon dioxide and sunlight differently; nitrogen may be obtained as salts in the soil or as free nitrogen in the air. Some animals require amino acids as components of their diets, others can build them from inorganic salts.

Chapter 4 puts our present efforts in a historical perspective, reviewing early processes of domestication and natural regions of variation. The first selections were subconscious, and the examples of wheat and maize are cited.

Growth of organisms is restricted by the supply of nutrients, and this forms the subject of Chapter 5. Plants are limited to the soil in which they grow, whereas animals can go and look for food. Even if food is available in the environment, the individual organism may have to fight competitors for it – other plants or animals. Humans can control the competitors, too, though this control may have side effects.

Two important facets of food production are growth rate and reproductive rate. Both are considered in Chapter 6, which also introduces some of the methods of increasing these rates, in micro-organisms, plants and animals.

Thus far we shall have been looking at the phenotypes of the individuals. Chapter 7 explains how these relate to the genotype, and what the one tells us about the other. It discusses some of the interactions of genes, and ways in which we can control the genes which pass to subsequent generations. Chapter 8 is about breeding systems, natural and artificial. To exploit these to our advantage, we must understand the implications of the genetics in the previous chapter. All this has been theory; Chapter 9 outlines real breeding programmes and their economics.

From the economic point of view, disease is a major threat to production. It is discussed in Chapter 10, as are forms of resistance to disease and control of disease.

Chapter 11 introduces the many uses of micro-organisms in producing or modifying our food. Such uses will grow in the future, and there may be other changes: we may find other sources of food, for example, or learn to use existing ones more effectively. Chapter 12 argues that such changes will depend on our greater understanding of the biology of the plants and animals which supply our food.

Summary

We have the responsibility to plan and prepare the use of natural resources for future generations.

We must consider both the benefits of efficient production of food for our ever-increasing world population and the idea that some areas should be maintained for social, recreational or aesthetic reasons.

Where we opt for maximum production of food, any or all of these options might be adopted:

a growing our existing organisms better than at present;

b breeding better, or more efficient, food-producing organisms;

c choosing the most appropriate organism in the food chain to utilise the *environment* and the *energy* from the sun (or fossil fuel) most effectively.

2 Plants or animals for food? Setting the scene

Humans are feeding organisms in a feeding environment. They are one of many types of animal and plant which exist together; some are predators, some are preyed upon, some are both.

This book is about the ways in which we may improve the growth of those organisms upon which we feed. To achieve this, we must know our own food requirements and the suitability of plants or animals; we can then make best use both of the food available already and of the natural capacity of the earth to produce more.

The food chain

Humans eat a variety of plant and animal foods but are themselves rarely eaten; they are at the end of many food chains.

The chains may be quite simple. For example, we eat plants, and are thus **primary** (1°) **consumers** since we feed on the **first trophic level** (as does the herbivore). But we also eat herbivores, such as cattle or sheep, and are therefore **secondary** (2°) **consumers**, in that we feed on the **second trophic level**. Because we do not usually eat large carnivores, there is rarely a further link in that chain; but we do often eat small carnivores such as fish, and so become 3° or 4° consumers.

Two principles of feeding

Two basic themes run through this chapter. The first is that *living plants and animals lose energy as heat to the environment*; the second is that *living plants and animals use components in their food to make more body tissue.*

When food passes along the food chain, there is an overall loss of energy; this principle is summarised by the **second law of thermo-**

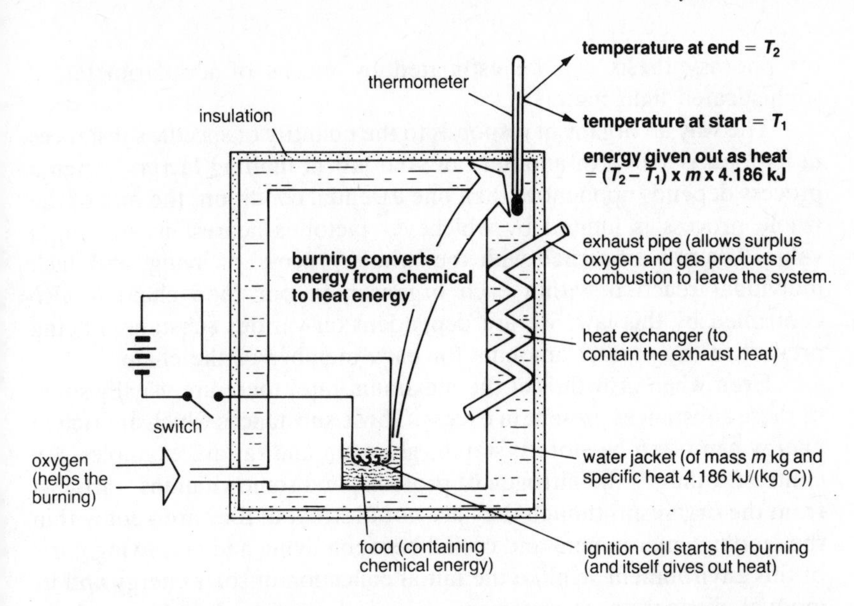

Figure 2.1 A bomb calorimeter. This apparatus measures the energy content of food. The food is ignited inside the insulated container, and the resulting rise in the temperature of the water jacket is used to estimate the heat released. The instrument measures energy content (not availability); coal, wood and wheat might yield the same heat, but we are able to use only one of these as a source of food energy.

dynamics, which states that for any reaction to proceed, there must be a net loss of heat energy. Because each of the many chemical reactions in any living organism produces heat, some chemical energy within the organism is lost and this means that there is a constant demand for energy for life and growth.

Efficient use of energy from the sun requires that this energy be intercepted or harvested with as little as possible being squandered as heat (either from the organism or as energy flows from one member of the food chain to the next). Plant food is thus the least wasteful way of feeding humans.

We can actually measure the energy content of food by burning it in a special container called a bomb calorimeter (Figure 2.1). The heat so formed, expressed in joules, is equivalent to the total chemical energy content of the food – though not all this energy is available to an animal which eats the food. Such measurements confirm that only a small proportion of the energy taken in by a herbivore (primary consumer) is retained and so made available to the secondary consumer.

Going back a step, we can also consider the incoming energy from the sun and the ability of the plant to collect and convert this into chemical energy. The amount of energy from the sun which is available for photosynthesis can be estimated by means of a solarimeter, a sophisticated light meter.

The way an organism responds to the quantity of specific substances in its food is stated in another law, the **law of limiting factors**: when a process depends upon more than one essential condition, the rate of the whole process is limited by whichever factor is nearest its minimum value. We are concerned with the 'whole process' of living, and each individual reaction within each organism in our food chain is also controlled by this law; we are dependent on various substances being present in the correct amounts for each member of the chain.

Even when growth is at the maximum rate, there are usually some of these substances present in excess. Those substances which are rich in energy can often be stored: starch, glycogen and fat are examples. By contrast, others – the nitrogen of proteins, and some vitamins – are lost from the organism, though they are not destroyed; they are kept within the feeding environment and cycled between living and non-living parts of this environment. Unlike the initial collection of solar energy and its gradual dissipation as organisms live and grow, substances such as hydrogen, phosphorus, sulphur and carbon are neither created nor lost; but the forms in which they occur may vary greatly.

Feeding is governed by these two principles of food production: the use of **nourishment** depends on the law of limiting factors; the use of **energy** depends on the second law of thermodynamics. An organism feeds to obtain nourishment and energy from its environment: 'nourishment' includes all the chemical building blocks which must be available for the organism's wellbeing; 'energy' is used to assemble these building blocks during growth, for reproduction and the maintenance of body structures, and in animals for movement.

Animal feeding

We can illustrate this rather general statement by looking at our own feeding. We need not only energy-yielding substances such as sugars, fats and starch, but also certain amino acids, vitamins, minerals and fatty acids; these are the building blocks. In practice, these two components of our food are rarely separate, and some building blocks supply energy also. (White sugar is an exception: an example of a food which generally supplies us with energy alone.)

Both types of food are used in our bodies to synthesise complex substances. Haemoglobin, for example, comprises building blocks such as amino acids, iron and parts of sugar molecules, and is assembled using the energy component of our diet.

The animals which we eat also need both types of food, but the types of building block may not be the same. Some animals, such as pigs and poultry, do need very similar substances; others, such as camels, cows and sheep can be fed on very simple chemicals which are often inorganic. In these animals, simple chemicals are gathered into complex substances not by the body tissues of the animal itself, but by micro-organisms living in its gut. We shall discuss these 'ruminants' in a later chapter.

Plant feeding

Humans, and their farm animals, need both complex building blocks and chemical energy; they are **heterotrophic** organisms. In contrast, green plants (for example) fix energy from sunlight and use it to make the sugars which are then used for growth; although some building blocks are still needed, these are simple inorganic substances which the plant gets from the soil and air (minerals such as nitrates, phosphates, potassium, calcium, magnesium and iron; plus water from the soil and carbon dioxide from the air). This is **autotrophic** nutrition. Chlorophyll, which gives plants their characteristic green colour, is an illustration of the way these substances are used; it is made of carbon, hydrogen, oxygen, nitrogen and magnesium, taken into the plant as inorganic substances and assembled using energy which came originally from the sun.

Our diet

Having mentioned the basic components of our diet, we now list in detail what is required for normal human growth.

An energy supply
This is provided by sugars, fats or starch, and to a limited extent by proteins.

A protein supply
This must include the so-called **essential amino acids**: leucine, isoleucine, tryptophan, valine, lysine, methionine, phenylalanine and threonine; children need arginine and histidine in addition.

Linoleic acid
This is an **essential fatty acid**, and comes from the fat in plant food.

Vitamins
A, B_1, B_2, B_6, B_{12}, C, D, E, K, pantothenic acid, biotin, folic acid and nicotinamide are provided by a wide range of plant and animal foods.

Minerals
These include calcium, phosphorus, potassium, magnesium, iron, sulphur, nitrogen, sodium and traces of many others.

Water
This is contained in most foods and in all our beverages.

Fibrous material
This is normally provided by the tough cell walls of plants.

Why do we need fats and carbohydrates?

There is an obvious need for energy when we move around. Active people need more food than those who take little exercise (Table 2.1), but there is also considerable variation among people; a fidgety person uses more energy than does a placid one, even if both have sedentary jobs.

As exercise is progressively reduced, the energy requirement decreases, but there is no sign of our energy consumption ever reaching zero – even when we sleep we use energy. This minimum consumption is the **basal metabolic rate (BMR)**. It varies between individuals, is lower for women than for men of similar weights, and is affected by certain hormones.

Energy is used all the time in ensuring that our vital organs continue to function, that proteins and other substances are synthesised, and that the nervous system is maintained in a state of readiness. Protein synthesis never stops; our tissues are continuously being renewed and this needs energy.

The second law of thermodynamics tells us that growth needs chemical energy. Growing children use more energy than their parents, in relation to body weight (Table 2.2); the energy content of food is used to assemble the new body tissue. Similarly, a pregnant woman needs more energy – energy equivalent to twenty 1-kg bags of sugar is used in making a baby. The nourishing of a baby with its mother's milk is also energy-intensive, equivalent to 175 g of sugar per day – much more demanding in terms of energy than making the baby in the first place.

Energy is needed for the actual digestion and absorption of food. This so-called **specific dynamic action (SDA)** or **thermic effect** of food is not large, about 5–10 per cent of BMR. The exact reasons for it are not clear. It is suggested that extra secretions are produced by the gut when we eat, and many of the absorptive processes need energy. Once the food is in the bloodstream, it must undergo chemical rearrangements; these **metabolic processes** occur in the liver, which account for its very large heat output.

We can store carbohydrates and fats. Carbohydrate is stored in the

Table 2.1 Energy in kilojoules used daily by average men and women (based on Taylor, 1978).

		8 h sleep	8 h work	8 h leisure	Total
Men	sedentary	2000	3500	5500	11 000
	moderately active	2000	5000	5500	12 500
	very active	2000	7500	5500	15 000
Women	office or home	1750	3500	3750	9 000

Table 2.2 Recommended daily energy intake (based on Taylor, 1978).

Age range	Energy intake (kJ)	Body weight (kg)	kJ/kg
0–3 months	2 300	4.6	500
9–12 months	4 200	9.5	440
4–5 years	7 000	17.4	400
Adult male	11 700	65	180

Table 2.3 Percentages of certain chemicals found in the body: the differences between men and women are due to the greater fat content of women; the other chemicals are in the same proportions in men and in women (based on Taylor, 1978).

	Men	Women
Water	62	54
Protein	17	15
Fat	14	25
Minerals	6	5
Carbohydrate	1	1

liver and muscles as glycogen which can be used over a short period. The liver can convert carbohydrates into fats for long-term storage. These fats, along with those derived from fat in the diet, are generally deposited under the skin.

In summary, our use of fats and carbohydrates provide energy for the following body functions: BMR, movement, dealing with food, making new tissues, and laying down stores of fat.

Why do we need proteins?

Much of our body tissue comprises protein (Table 2.3), which is one reason why we need protein in our diet. Protein in food is broken down in the gut to its constituent amino acids, and these units are absorbed into the capillaries of the villi, pass along the hepatic portal vein and reach the liver where the amino (nitrogen-containing) groups are transferred to carbon-containing compounds to form new amino acids when necessary; this is called **transamination**. The whole mixture is passed round the body and the amino acids are used by the cells to make proteins.

We cannot store amino acids, though the reason for this is not known. Because we cannot store them, any which are not used to make body proteins must be disposed of; in **deamination**, the amino group is removed (to form urea) and an energy-containing carbon compound is left, which can generally be used for cellular respiration. The free amino group readily forms ammonia, but this is very poisonous to our cells. A cyclical series of reactions, the **ornithine cycle**, is responsible for converting the poisonous ammonia to urea. Even this, which is relatively safe, must be removed in solution as urine.

When the body is making a lot of tissue protein, as when young people grow, the use of amino acids by the body is high; and we may expect children to have a greater need for protein in their diet than do their parents (Table 2.4).

Amino acids are used efficiently only if they are present in blood in the same proportions as required for the protein under construction. Some amino acids can be made by the process of transamination, provided that the total protein intake is adequate; these are the so-called **non-essential amino acids**. The proportions of **essential amino acids** in the blood are identical to their proportions in the food. Whichever amino acid is present in the lowest proportion limits the rate of protein synthesis; low levels in food of just one amino acid can impose on young people a slow growth rate. The law of limiting factors applies to protein synthesis as it does to nutrition as a whole.

Table 2.4 Recommended intake of high-quality protein per day (based on Taylor, 1978).

Age	Protein intake (g/person)	Protein intake (g/kg body weight)
6–11 months	14	1.53
4–6 years	20	1.01
male 10–12 years	30	0.81
female 10–12 years	29	0.76
male 16–19 years	38	0.60
female 16–19 years	30	0.55
male adult	37	0.57
female adult	29	0.52

Protein balance

As shown in Table 2.4, adults (who are not growing) need proteins. This is because the body is inefficient in two ways; firstly, cellular structures cease to function normally, and are then broken down by enzymes; secondly, only 80 per cent of the amino acids released are re-used in replacing the structures, the rest being deaminated and lost. In this way, an adult loses nitrogen equivalent to 30 g of protein per day.

Inefficient re-use of tissue protein is the major reason for our continued protein requirement, but small quantities are also used to make proteins which are lost from the body, skin, hair, nails, certain hormones, and proteins in milk and babies. (A pregnant or nursing woman has a high requirement for protein as well as for energy.)

When the amount of nitrogen in dietary protein exactly matches the amount lost from the body, there is a state of **nitrogen balance**. On the other hand, growth results in a positive protein balance (a net gain) and conditions of protein starvation result in a negative protein balance (a net loss) because the body cannot completely re-use the amino acids from proteins in dead tissues.

A negative protein balance may also result when amino acids in the food are in the wrong proportions. Nitrogen from amino acids present in quantities higher than needed is lost from the body when just one essential amino acid limits protein synthesis. An example is the body's use of the amino acids in gelatin. This protein (used in table jelly) lacks tryptophan and sulphur-containing amino acids, so no matter how much of this protein we ate, there would always be a negative protein balance if this was our only supply of amino acids. The total loss of nitrogen would be equivalent to the quantity of gelatin in the diet plus 30 g protein equivalent daily loss.

Why do we need vitamins and essential fatty acids?

Humans, like some other heterotrophic organisms, need vitamins and essential fatty acids in food. Whereas we need each day some grams of protein, milligrams only of vitamins are needed. Vitamins are small molecules (Figure 2.2) and were once thought to be 'vital amines'; like amino acids, few are stored in the body, so regular supplies are needed. Most good standard biology textbooks discuss the many different vitamins which we need; one example will illustrate their importance in our diet.

It is well known that vitamin C deficiency is associated with **scurvy**, a disease of sailors when boat journeys took a long time and food conservation was limited. The characteristic symptoms are that the gums bleed and in severe cases wounds fail to heal. Vitamin C is known

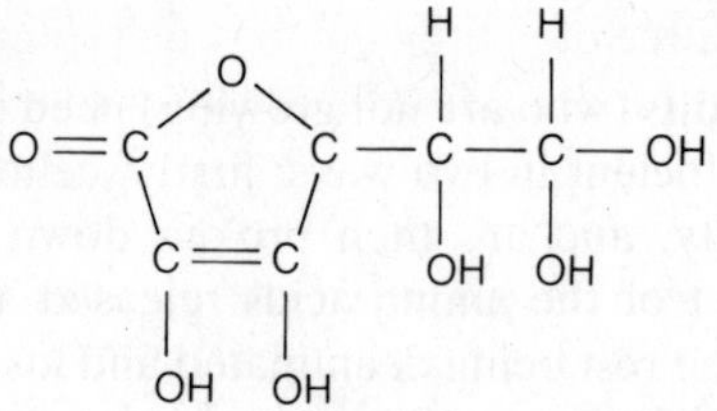

H H
O
$O{=}C$ C — C — C — OH
$C{=}C$ OH OH
OH OH

Figure 2.2 The molecular structure of vitamin C (ascorbic acid). The molecule has six carbon atoms, and is related to hexose sugars.

C18:2	$CH_3(CH_2)_4CH{=}CHCH_2CH{=}CH(CH_2)_7COOH$	**linoleic**
C18:3	$CH_3CH_2CH{=}CHCH_2CH{=}CHCH_2CH{=}CH(CH_2)_7COOH$	**linolenic**
C20:6	$CH_3(CH_2)_4(CH{=}CHCH_2)_3CH{=}CH(CH_2)_3COOH$	**arachidonic**

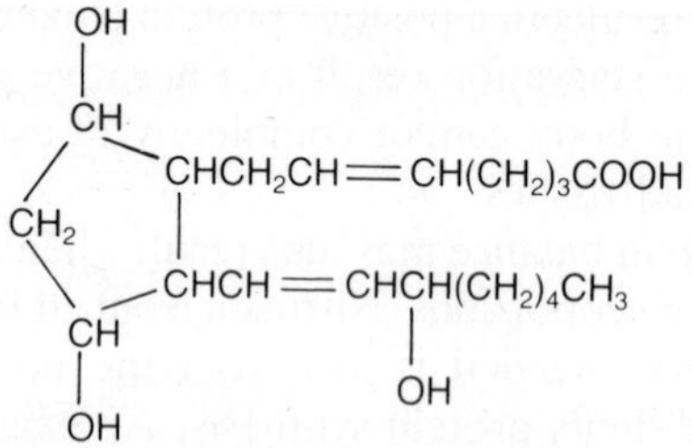

OH
CH
$CHCH_2CH{=}CH(CH_2)_3COOH$
CH_2
$CHCH{=}CHCH(CH_2)_4CH_3$
CH
OH
OH

prostaglandin $PGF_{2\alpha}$

C20:2

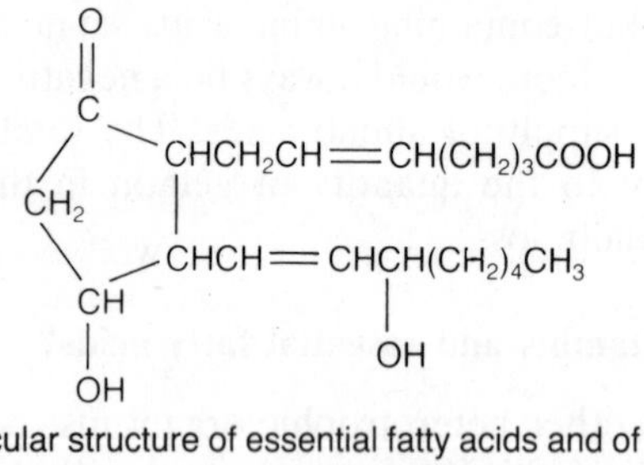

O
C
$CHCH_2CH{=}CH(CH_2)_3COOH$
CH_2
$CHCH{=}CHCH(CH_2)_4CH_3$
CH
OH
OH

prostaglandin PGE_2

Figure 2.3 The molecular structure of essential fatty acids and of a prostaglandin derived from these.

to be involved in the synthesis of collagen (fibres of which make up connective tissue) and collagen is involved in wound healing, so collagen may link the symptoms of deficiency with the biochemical function of the vitamin.

About 10 mg of vitamin C is needed each day to prevent scurvy, but 50–60 mg is recommended. Nobel Laureate Linus Pauling has suggested that an intake of 2000 mg per day may reduce both the chance of catching the common cold and its duration if caught; the role of vitamin C here is unknown.

Fats are a very rich source of energy but may be replaced in the diet by other energy-yielding foods. Many cellular structures (such as membranes) are made from fats; and certain hormones (such as prostaglandins) are based on fatty acids. We cannot make at least one fatty acid, linoleic acid, which has eighteen carbon atoms (Figure 2.3); this is thus an essential fatty acid – it must be provided in the diet. Two others are beneficial to growth: linolenic acid and arachidonic acid. Linoleic is probably converted into linolenic and then to arachidonic acid, these then being used by the body to make other chemicals.

Fat in the diet helps in the absorption of fat-soluble vitamins such as A and D.

Plants and animals as food makers

Our first principle of feeding – the second law of thermodynamics – dictates that only some of the light energy trapped by the leaves of a green plant is passed on to the flesh of a herbivore. Energy is again lost when we eat meat.

Of the solar energy which strikes plants in a field, only 1–5 per cent is used for photosynthesis. The conversion by herbivores of plant material into meat is 5–20 per cent efficient, energy again being lost as heat. Carnivores, similarly, can utilise herbivore flesh with 5–20 per cent efficiency. We can choose to grow plants and animals for food which have higher rather than lower efficiencies, but the energy losses are still considerable.

So far as our food production is concerned, a given area of land can produce more food if it is used to grow plants rather than animals. Calculations indicate that one hectare of well-managed wheat can support about fourteen people if they use only grain for food; if they use animal products for half their energy intake, that hectare can support only four or five people. Under conditions of shortage (and provided that wheat is a satisfactory food), we should therefore eat the wheat itself, rather than convert part of it into meat. But plant foods are not entirely satisfactory, and so there are advantages in including animal products in our diet.

Plants and animals meeting our daily needs

Animal products are characterised by their high protein content. Dairy products provide fat-soluble vitamins, calcium, protein and some fat. Plant products are rich either in energy-yielding compounds (plants such as cereals, potatoes and yams) or in minerals and water-soluble vitamins (vegetables and fruits). Our basic food requirements of protein and energy can therefore be met most effectively by animal products and plant products respectively, but a suitable mixture of the so-called 'protective foods' is also required to supply traces of vitamins and minerals. When there are no food shortages, we should eat a wide spectrum of foods to provide a balanced diet.

Interpretation of the second law of thermodynamics implies that we may be obliged to eat plants because of their higher food potential per unit of land. Yet in strictly nutritional terms, plant protein is inadequate in both quantity and quality, so that the less-productive animals have their advantages.

Plant protein quantity

An average adult requires about 30–40 g of high-quality protein per day along with 11 000–12 500 kJ per day. This is equivalent to a **protein/energy (P/E) ratio** of 2.5–3.5 mg protein/kJ; using a staple food with a ratio below this results in protein malnutrition. Fortunately, few plants are so poor in protein (Table 2.5). However, as we shall see, the quality of their protein may reduce the use our bodies make of them.

Table 2.5 The protein/joule (P/J) ratio of selected foods based upon protein and energy content of 100 g of food. The *minimum* P/J ratios for infants and sedentary males indicate the vulnerability of young to protein malnutrition (kwashiorkor) (based on various sources).

Food	Protein (g) in 100 g of food	Energy (kJ) in 100 g of food	P/J ratio (g protein per 100 kJ)
fish	16.0	289	5.54
meat	14.8	1311	1.13
milk	3.3	272	1.21
beans	7.2	289	2.49
peanuts	28.1	2455	1.14
wheat	11.6	1433	0.81
potatoes	2.1	318	0.66
maize	7.5	1529	0.49
rice	6.2	1504	0.41
banana	1.1	310	0.35

Minimum P/J ratio for human infant is 0.57
Minimum P/J ratio for human adult is 0.36

Malnutrition

In some parts of the world, malnutrition among adults is a major problem, but it is among young children and pregnant and nursing mothers that these problems are greatest. These individuals are manufacturing large quantities of proteins and so require large quantities of dietary protein and energy. They are also the members of society least capable of ensuring that they obtain their fair share of the food available. When a low P/E food is used, the body may not be able to carry out tissue repair and protein growth. Prolonged negative nitrogen balance leads to protein malnutrition or **kwashiorkor**.

Kwashiorkor generally develops in young children who are weaned onto a low-protein diet which contains adequate quantities of energy, a situation summarised by a translation of the word 'kwashiorkor': 'the disease a child gets when another baby is born'. The characteristic pot belly is made more noticeable by wasted arm and leg muscles; the child is stunted, susceptible to infection and will eventually die unless the diet is corrected.

This should not be confused with general malnutrition, called **marasmus**. Again there is a negative nitrogen balance, but this is because tissue protein is broken down to meet the body's energy need;

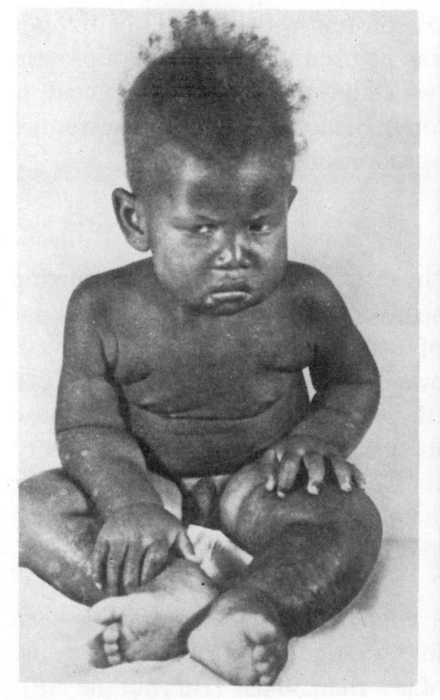

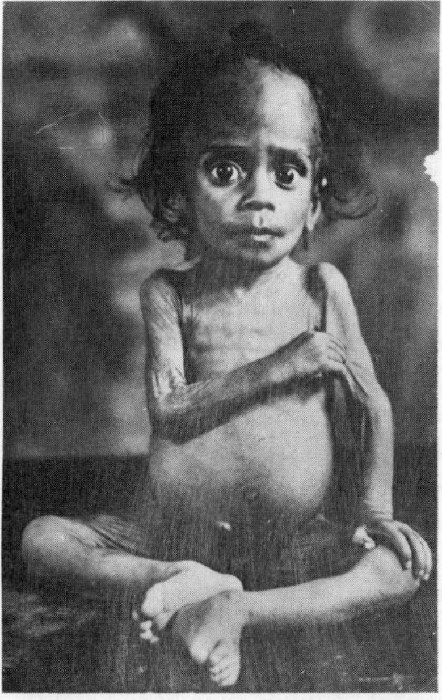

Figure 2.4 Kwashiorkor (left) and marasmus (right).

amino acids from dietary and tissue protein are used in cellular respiration, with the nitrogen being lost as urea.

Nutritional marasmus is common when infants are weaned early in life. The nutritious diet of mother's milk is replaced by a low-protein, low-energy diet. The body wastes away to meet the basic needs of keeping it alive (so there is no pot belly), the child becomes stunted, often suffers permanent brain damage and is liable to infection. Figure 2.4 shows the symptoms of marasmus and of kwashiorkor.

Nutritional marasmus can be countered by increases in energy-rich foods and in usable protein in the food; kwashiorkor by foods containing more usable protein.

The quality of plant proteins

The ratio of the essential amino acids in a protein largely governs the overall use of that protein. One of the proteins most efficiently used by the body is egg white; gelatin, on the other hand, cannot be used at all if it is the sole source of protein. The body's use of any protein can be compared with its use of a **reference protein**, usually egg white, which we regard as having a **net protein utilisation (NPU)** of 100 per cent. NPUs of less than 100 per cent reflect the imbalance of essential amino acids and show the percentage of nitrogen in the protein which can be incorporated into body proteins. So a protein with an NPU of 100 per cent, a so-called **first-class protein**, can be used with no waste of protein nitrogen; while groundnut protein, which has an NPU of 48 per cent, is a **second-class protein**, half of the protein being wasted because essential amino acids are not in the correct proportions.

Protein complementation

The low NPUs of plant proteins can be dramatically improved by mixing or **complementing** foods. While the total protein nitrogen is not increased in the diet, the proportion of essential amino acids is improved. So a food containing protein known to be deficient in one or two essential amino acids is mixed with protein especially rich in those particular essential amino acids, and vice versa. The dramatic effect of complementation is shown when rats are fed gelatin and wheat protein (Figure 2.5). Gelatin alone results in a negative nitrogen balance, which accounts for a steady decline in rat body weight; when wheat is added, the rats grow normally.

Another, more practical application of complementation is seen when we mix two common proteins in equal proportions. Bean and wheat proteins each have an NPU of less than 50 per cent, but the NPU of a 50 : 50 mixture exceeds 75 per cent. Beans contain large quantities of lysine and small quantities of sulphur-containing amino acids, whereas

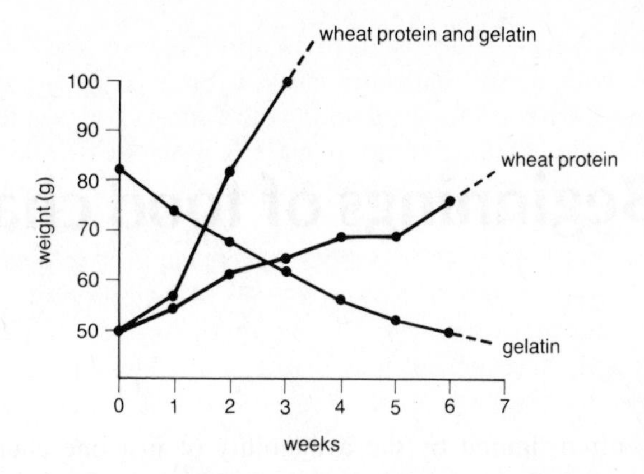

Figure 2.5 The effects of feeding rats gelatin, wheat, or a mixture of the two.

wheat contains small quantities of lysine and large quantities of sulphur-containing amino acids.

These are just two instances of protein complementation; cereals (wheat, rice or maize) may be mixed with legumes (beans, peas or soya beans), and legumes may be mixed with seeds (sunflower or sesame). The establishment of all of these combinations requires knowledge of the composition of foodstuffs as well as details of our own amino acid requirements.

The conflict

Plant foods have a much greater potential to support humans, yet because their building blocks differ in proportion to our own, they do not form ideal foods. When fed as staple foods, plant tissues can result in nutritional disorders. On the other hand, we see the enormous potential which plants have to alleviate hunger. Agriculturalists and nutritionalists must together identify organisms which utilise the resources of our planet to meet fully our dietary requirements.

Summary

Humans, like any other plants or animals, have certain nutritional needs which can be identified. In general, a varied diet is favoured, but our bodies use animal tissues more effectively than those from plants. Plants are more efficient producers of biomass which is rich in carbohydrate, but the protein is of low quality and quantity. The value of plant food may be improved:

a by complementing proteins of low quality;

b by cultivation of plants which have higher, rather than lower, protein quality and quantity.

3 Beginnings of food chains

Growth is often limited by the availability of just one chemical in a metabolic pathway, so the value of an organism as food depends not only on its quality as a food but also on its ease of growth. Closely related food producers may grow at slightly different rates: we grow for food the organism with the greater rate.

In contrast to these small variations in breeding or strain type, some organisms have fundamentally different ways of making food. These alternative ways of making food employ different biochemical pathways depending on how the organism gathers the building blocks from its environment. Two such building blocks are carbon dioxide and nitrogen: we shall consider two ways in which carbon dioxide may enter plants, and two alternative ways in which nitrogen may enter both plants and animals. These pathways have evolved to meet the specific needs of the organism, so we must understand them in order to grow the appropriate organism in a particular environment.

Origins of carbohydrate food

In order to make carbohydrate-rich foods, the plant has to do two things. Firstly, the carbon dioxide of the atmosphere must be chemically trapped within the plant. Secondly, it must be chemically reduced: carbon in carbon dioxide (CO_2) is highly oxidised, while in carbohydrate $(CH_2O)_n$ it is less oxidised (that is, more reduced). There are two ways of carrying out these processes.

Converting carbon dioxide into plant material

One way of measuring a crop's rate of photosynthesis is to record the increase in dry weight of crop in 1 m^2 of soil during a certain period of time. This increase is the **net primary production (NPP)**. The word 'net' is used because some of the sugar formed through photosynthesis is

Table 3.1 The growth of maize and sugar beet under different conditions (based on various sources).

Plant	Region	Net primary production ($g\ m^{-2}\ day^{-1}$)	Irradiance ($J\ cm^{-2}\ day^{-1}$)	Conversion of photosynthetically active radiation (%)
sugar beet	California	8	2000	1.4
sugar beet	Europe	31	1200	9.5
maize	California	52	2000	9.8
maize	Europe	17	1200	4.6

dissipated by the plants' own night-time respiration. Thus the dry weight actually measured is less than the total made, the **gross primary production (GPP)**. When a plant is not growing, its NPP is obviously zero; but in wild and cultivated plants we find a complete range up to 50 g m^{-2} day^{-1}. Table 3.1 shows the highest possible values which we can get for sugar beet and maize. The scientists who obtained these figures ensured that there was no wasted light, and that water and soil minerals were all present in adequate quantities. (The discussion in this section assumes that neither water nor minerals are restricting growth.)

An important point emerges from these figures. The European climate, cool and not very sunny, favours the growth of beet; while the hot, sunny Californian climate favours maize.

We can confirm this finding in the laboratory, by measuring the net rate of photosynthesis when a leaf is held in a special chamber at various temperatures or light intensities. The chamber is supplied with air

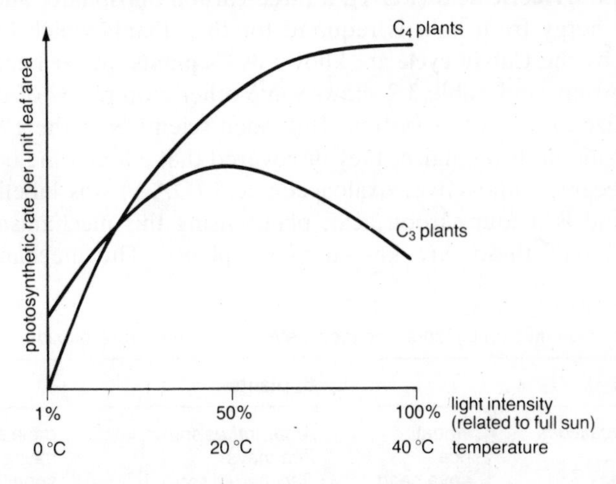

Figure 3.1 The effect of increased light intensity and temperature on the photosynthetic rates of C_3 and C_4 plants.

whose temperature, carbon dioxide level and humidity are known and steady. The gas is continually sampled as it leaves the chamber, and the carbon dioxide level measured by passing the gas through an instrument called an infra-red gas analyser (IRGA), which utilises the fact that carbon dioxide absorbs infra-red light.

The graph of photosynthesis by maize and sugar beet at varying temperatures and varying light intensities, confirms our original conclusions (Figure 3.1).

To explain these observations on the nature of the incorporation of carbon dioxide into plant material, we must turn to the biochemical pathways of photosynthesis which are found in maize and sugar beet.

Two types of photosynthesis

Sugar beet and maize typify two groups of plants with very different methods of incorporating carbon dioxide into plant organic material. The two ways account for the bulk of carbon dioxide fixed by plants.

Let us look first at the basic **carbon fixation cycle** which was worked out by Melvin Calvin in the early 1950s. He employed the then new technique of **radioactive labelling**. Carbon dioxide was 'labelled' with carbon-14 isotope. After the photosynthesising plant tissue had been exposed to this for a very short time, the tissue was killed (by dropping it into boiling ethanol) to prevent further enzyme action. Calvin then separated the chemicals in the tissue (using paper chromatography) and identified them. By taking samples over a five-minute period at intervals of a few seconds, he was able to show that radioactivity was found first in phosphoglyceric acid (PGA), a three-carbon phosphate, and then in sugar. Energy from light is required for this. Plants which fix carbon dioxide by the **Calvin cycle** are known as **C_3 plants**. Sugar beet belongs to this group, and Table 3.2 shows some other crop plants of this type.

Maize also has this system. But when scientists in the 1960s used Calvin's methods on maize, they discovered that a long time before the PGA became radioactive, oxaloacetic acid (OAA) was labelled. This compound is a four-carbon acid; plants using this mechanism, maize being one of these, are known as **C_4 plants**. The mechanism was

Table 3.2 Common crop plants, and their methods of fixing carbon dioxide.

C_3 plants		**C_4 plants**	
Triticum aestivum	wheat	*Amaranthus* spp.	grain amaranth
Oryza spp.	rice	*Zea mays*	maize
Glycine max	soya bean	*Saccharum* spp.	sugar cane
Hordeum vulgare	barley	*Sorghum bicolor*	sorghum
Solanum tuberosum	potato	*Pennisetum purpureum*	elephant-grass

elucidated by Hal Hatch and Roger Slack, and is called the **Hatch and Slack cycle**.

Having identified the chemicals in any new cycle and understood the sequence of their involvement in it, biochemists try to isolate and characterise each enzyme involved. This has been achieved for all reactions in each of the cycles mentioned above; of the very many individual enzymes, the two which have a profound effect upon agricultural productivity are those which gather carbon dioxide from the air.

Enzymes which trap carbon dioxide

The enzyme which gathers carbon dioxide in the C_3 (Calvin cycle) plant is ribulose diphosphate carboxylase, so called because carbon dioxide is added to ribulose diphosphate which then breaks down to PGA. In the C_4 plant (Hatch and Slack pathway), phosphoenol pyruvate (PEP) carboxylase adds carbon dioxide to PEP to form OAA. We now know that the C_3 system can gather carbon dioxide less well than the C_4 system. This is shown by determining the 'compensation point' of the C_3 and C_4 plants.

We use an experimental set-up similar to that used to determine the photosynthetic rates, but now the system is closed: air in the growing compartment is sampled by an IRGA and then returned to the compartment. When the leaf is illuminated, carbon dioxide is removed from the air; and we find that the C_3 plant, sugar beet, removes carbon dioxide until the air has fallen from the normal value of 300 parts per million (ppm) (0.03 per cent) to about 50 ppm. The C_4 plant, maize, on the other hand, can reduce the carbon dioxide level to about 1 ppm. Because a state of equilibrium exists, we call these levels **compensation points**. If air is introduced into the chamber with carbon dioxide at a level lower than the plant's compensation point, the plant will contribute carbon dioxide to the air by respiration until the level reaches the compensation point, even though all other conditions favour photosynthesis. The C_4 enzyme is so good at extracting carbon dioxide from the air that when both maize and sugar beet are grown with limited carbon dioxide in the same sealed growing compartment, the maize actually fixes carbon dioxide released by the sugar beet!

The increase in efficiency of photosynthesis in the C_4 plant can be accounted for by the enzyme PEP carboxylase. It is particularly valuable to the plant when carbon dioxide levels are low and photosynthesis is taking place rapidly under conditions of high light intensity: conditions found in the tropics. The decline in efficiency of the C_3 plant as intensity and temperature increase is caused by a phenomenon known as photorespiration.

Photorespiration and net rates of photosynthesis

Photorespiration is the production of carbon dioxide under conditions which favour photosynthesis; it seems as though the high levels of oxygen produced by active photosynthesis stimulate the production of carbon dioxide by a mechanism distinct from normal respiration.

Only C_3 plants carry out photorespiration; these plants squander a lot of the sugar formed by photosynthesis (**gross photosynthesis**) leaving less for growth (**net photosynthesis**). By contrast it seems as though the C_4 plant uses more of its gross photosynthate for growth. This lack of photorespiration in C_4 plants can also account for the high levels of net primary production, shown in Table 3.1, for maize grown in California.

The link between the high-affinity enzyme and lack of photorespiration in the C_4 plant is unclear, but it seems that gross rates of photosynthesis in C_3 plants are just as high under natural conditions as are rates for C_4 plants, so we now think that it is the level of photorespiration which largely determines net primary production.

The exciting possibility exists that the agriculturalist and the biochemist together might reduce or totally block photorespiration: chemicals are already known which inhibit this process. Experimentally, if these are sprayed into crops, net productivity can double, showing that up to half the carbon dioxide fixed by photosynthesis is normally lost by photorespiration. Perhaps scientists will eventually establish genetic control over these biochemical processes and then modify the pathways in certain crops.

Why are there any C_3 plants?

If the C_3 method of carbon dioxide fixation is relatively so inefficient, why have C_4 plants not swept to supremacy? There are three reasons. Firstly, C_3 plants perform better than C_4 plants under temperate conditions because lower light intensities and temperatures do not favour photorespiration (Figure 3.1 and Table 3.1). Indeed, increased yields of wheat (a C_3 plant) in the USA over the last 50 years can be attributed in part to the early maturity of new varieties, which enables the grain to fill before the high temperatures of summer stimulate photorespiration and thereby dissipate the gross photosynthesis; this occurs in older, later-maturing varieties. However, true comparisons between C_3 and C_4 plants are difficult because the other characteristics shown by the plants are so very different; the difference between C_3 and C_4 plants under temperate conditions may stem from this 'genetic background'. As an extreme example, frost-hardy temperate grasses growing in Europe may accumulate 1 kg m^{-2} of dry matter before the frost-tender maize can be planted. When the genetic background is

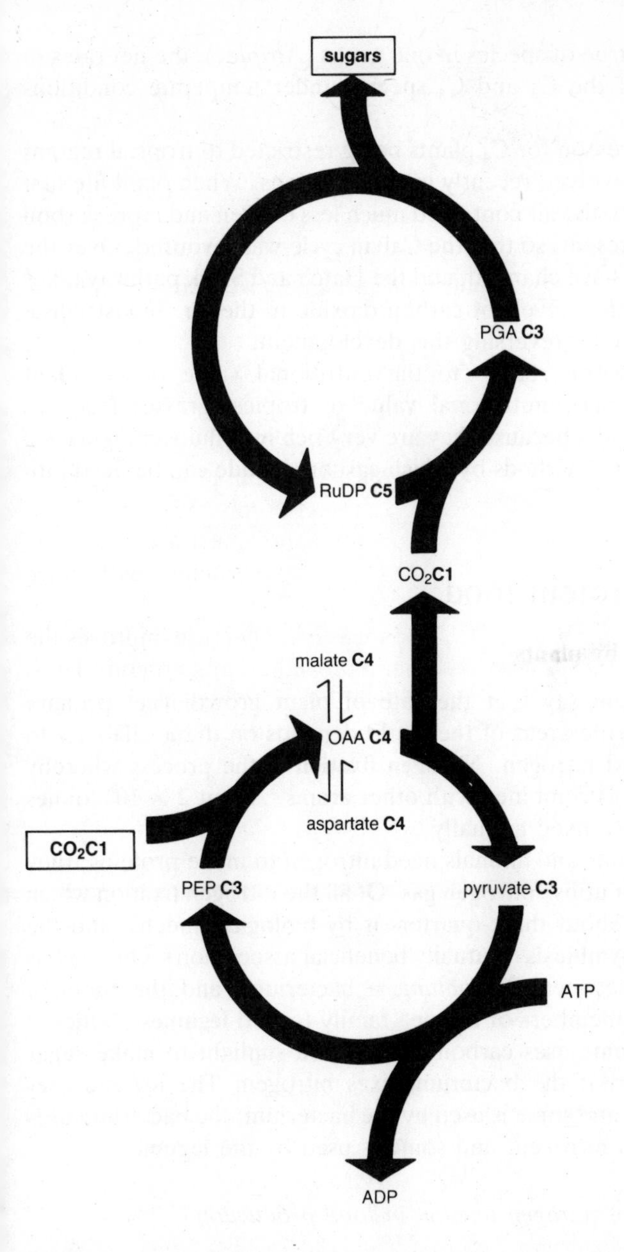

Figure 3.2 Two methods of carbon dioxide fixation used by plants (see text for abbreviations used).

quite close, as is true of species in one genus (*Atriplex*), the net rates of photosynthesis of the C_3 and C_4 species under temperate conditions differ little.

The second reason for C_4 plants being restricted to tropical regions is that they have evolved recently in those regions. When plant life first appeared on earth, the air contained much less oxygen and more carbon dioxide than at present, so that the Calvin cycle was favoured. Over the years these levels have changed, and the Hatch and Slack pathway takes advantage of the low levels of carbon dioxide in the air; industrialisation, though, may be reversing this development.

The third reason relates to the nutritional value of the plant material formed. The nutritional value of tropical grasses (such as elephant grass) is less because they are very rich in cellulose. Figure 3.2 summarises the two methods by which carbon dioxide can be fixed into plant material.

Origins of protein food

Nitrogen fixation by plants

In general, we can say that the rate of plant growth (net primary production) in fertile areas of the world depends on the availability to the plants of fixed nitrogen. **Nitrogen fixation** is the process whereby nitrogen gas (N_2) is combined with other atoms. About 2×10^8 tonnes of nitrogen gas are fixed annually.

Although plants and animals need nitrogen to make proteins, they themselves cannot utilise nitrogen gas. Of all the nitrogen fixation which does take place, about three-quarters is by biological means; and the bulk of this is by **symbiosis** (mutually beneficial association). One such is the association between *Rhizobium*, a bacterium, and the roots of plants, generally members of the pea family (called **legumes**), which it invades. The legume uses carbon dioxide and sunlight to make sugar (via photosynthesis); the bacterium fixes nitrogen. The legume uses some of its sugar, and some is used by the bacterium; the bacterium uses some of the fixed nitrogen, and some is used by the legume.

The importance of nitrogen fixation in food production

While we can increase nitrogen supply to crops by means of this association, the benefit to us relies upon the use we can make of leguminous crops – pulses, clover, alfalfa and others. Like other plants, they can be eaten by animals. Their protein-rich seeds (peas, beans and soya beans) can be used in our food. Because of the advantage in using plant foods, soya protein is much used in the manufacture of artificial meat.

Another way of exploiting this association is to grow the legume to enrich the soil; after a short period of growth it is ploughed into the land. This can be done either between existing crop plants, as in an orchard, or prior to planting a crop which does not fix nitrogen; the process is called **green manuring**. Alternatively, the soil can be enriched by **crop rotation:** nitrogen fixation is stimulated by low levels of fixed nitrogen in the soil, whereas high levels of nitrogen, which develop when legumes are grown over a long period of time, inhibit more fixation. Some countries have exploited this to the full; in Australia, 99 per cent of the nitrogen fixation is effected by the legume–*Rhizobium* relationship. By contrast, this fixation accounts for about half the total in India, the rest being supplied by nitrogen fertiliser; in fact, India uses almost ten times the quantity of nitrogenous fertilisers used by Australia. Every country should seek to maximise biological fixation; the improvements in yields and conservation of natural resources are enormous.

The mechanism of fixation

Details of the fixation process are complex, but it seems that the enzyme which fixes nitrogen, nitrogenase, comprises two sub-units which combine to transfer hydrogen to the nitrogen gas. The enzyme seems to be similar irrespective of its source. So the same basic system of nitrogen fixation is shared by *Rhizobium*, by other non-symbiotic bacteria (often anaerobic) and by the blue-green algae.

Oxygen gas is known to inhibit the enzyme, and it is now thought that a special oxygen carrier ensures that *Rhizobium* bacteria get oxygen for respiration while anaerobic conditions are maintained for the enzyme. The legume produces this carrier; called **leghaemoglobin**, it is very similar to haemoglobin and makes the nodules pink.

*Are there any drawbacks in the legume–*Rhizobium *association?*

An environmentalist might argue that if a weed had the ability to fix nitrogen and if this weed grew prolifically, it would enhance the nitrogen status of the soil to such an extent that water supplies could be polluted by nitrate leached from the soil. Under anaerobic conditions, nitrate may be partially reduced to the very poisonous nitrite; the resulting problems are well known as a consequence of extensive application of nitrogen fertiliser. But it seems unlikely that we need concern ourselves with the possibility of a rampaging nitrogen-fixing weed!

The most relevant drawback is a structural property of the legume.

In grasses, until they flower, the growing point remains very low and so the plants can be eaten, mown and trampled by stock, with little reduction of their capacity to regenerate new leaves and grow. Legumes, however, have growing points at the end of the growing shoot, and are readily damaged. When grass and legumes are grown together, the legume can easily be lost from the pasture unless particular care is taken by the farmer.

Additionally, grasses grow more rapidly than legumes when fixed nitrogen is available in the soil. This is because the reactions in the legumes which fix nitrogen gas all require energy, and so less energy is available for growth. Yields of a legume grain are generally half those of a cereal grain.

Nevertheless, enormous contributions can and must be made by legumes to the fixation of nitrogen. When food and natural resources are scarce, this feature of the legume–*Rhizobium* symbiotic relationship must be further exploited.

'Organically grown' plants – a digression

We have seen that some plants can fix nitrogen by their symbiotic relationship with *Rhizobium*, while others rely entirely upon the levels of fixed nitrogen in the soil. These supplies of soil nitrogen may come from the organic remains of plants and animals, or they may be added to the soil as inorganic salts of nitrogen in fertilisers used by the farmer.

Some people feel that food grown using intensive, high-yielding methods is inferior to that grown in a less intensive system. This has led to the view that organically grown crops – cereals for instance – are preferable to those grown using inorganic fertilisers. But consider for a moment the plant and its feeding environment: by definition, an autotrophic organism takes up simple inorganic substances: does it matter to the plant where these minerals in the soil come from?

Organic or inorganic nitrogen for animals?

Both plants and animals feed on fixed nitrogen. Plants use inorganic nitrogen (NH_4^+ or NO_3^-) and we generally assume that all animals require some essential amino acids or at least the nitrogen in combination with an organic compound. However, this is not so. Cows, sheep, deer, goats and camels all belong to a group of animals which can use ammonia and other sources of inorganic nitrogen, and do not rely on organic nitrogen. They are called **ruminants**. Micro-organisms in their stomachs use the inorganic nitrogen to build up their own protein; these in turn are digested and absorbed by the host animals. In this way, animals cease to be dependent for protein nitrogen on plants or other free-living animals.

Under natural conditions, a diet of inorganic nitrogen is unlikely, but ruminant digestion ensures that growth is never restricted by the lack of just one essential amino acid – as is sometimes the case with humans – since micro-organisms have enzymes which make all essential amino acids. So a ruminant could grow and thrive, even if fed sufficient gelatin (which contains no tryptophan) as its sole dietary source of nitrogen (see page 19).

Ruminant digestion – a closer look

Micro-organisms grow in a special region of the stomach known as the **rumen** (see Figure 3.3). When the animal takes grass, this is chewed only slightly before being passed to the rumen, where microbial action takes place, aided by the neutral pH and by further chewing after regurgitation (chewing the cud). The microbial action, or **fermentation**, breaks down the vegetation, including the cellulose cell walls, and the micro-organisms multiply. So the bulk of the animal's dietary vegetation is incorporated first into micro-organisms, and at the same time large quantities of various organic acids are produced. What then are the advantages of fermentation?

Firstly, as we have seen, ruminants do not have an 'essential' amino acid requirement. Because of the micro-organisms' abilities to use inorganic nitrogen and to break down cellulose, ruminants can be fed on unconventional foods. In theory, a cow could obtain its food requirements from straw or even waste paper, and from a supply of fixed nitrogen (urea or ammonium ions, for example). In practice, the bulkiness of the straw and the toxicity of the ammonium ions might present problems. Nevertheless, cellulose can constitute an energy-rich component of the diet rather than an inert, bulky substance, and in a heterotrophic higher animal, ammonium ions can be used as building blocks. So the straw content of cattle feed can be raised in some situations to one-third without loss of production.

Secondly, the volatile fatty acids (VFAs) produced during fermentation are absorbed by the gut, and it is these energy-rich compounds rather than glucose which provide the animal with the bulk of its energy needs – blood-sugar levels are conspicuously low.

Some fermentation products are lost – methane, for example, is produced in this way and burped (eructed) out, thus losing about one-fifth of the energy intake.

The essential requirement of microbially supplemented digestion is that the food must be exposed to these symbionts for a long period of time. The true ruminant does this by retaining the food in a large fermentation chamber (the rumen) before it enters the true stomach, a process called **pre-gastric fermentation**. This contrasts with the **post-gastric fermentation** in such herbivores as the horse, where the VFAs are produced in the large intestine (Figure 3.3); the horse has a particularly well-developed large intestine which allows this fermentation to take place.

The rabbit slows down the passage of food to an enormous extent to allow microbial activity by eating its own faeces. Clearly, this process

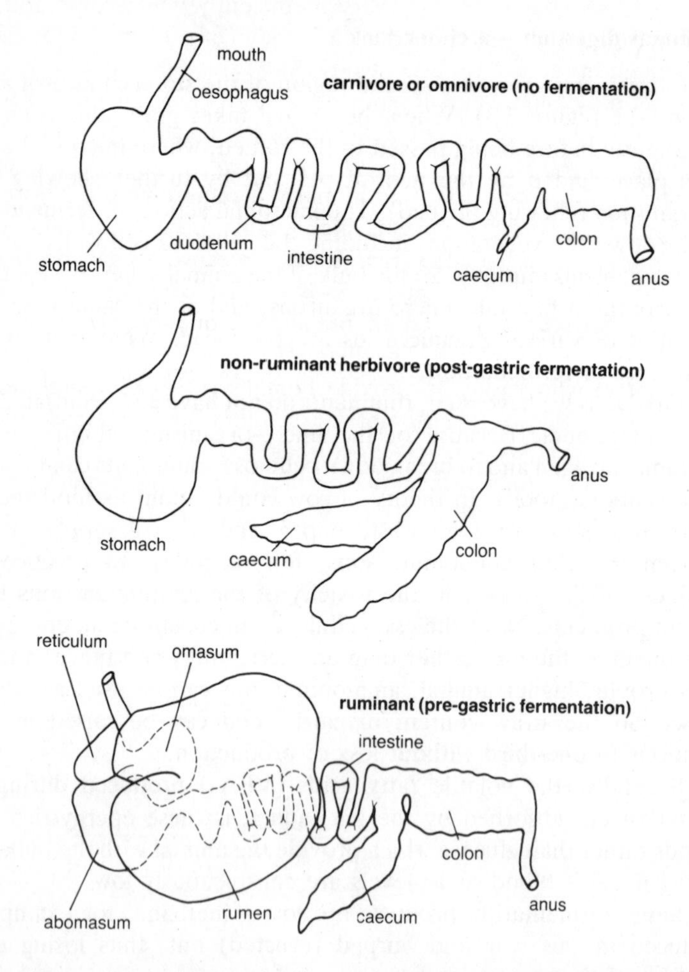

Figure 3.3 Three basic patterns of mammalian gut; structure and function are closely related. In the omnivore, there is little fermentation. In the ruminant herbivore, food is fermented and cellulose broken down before it reaches the stomach (pre-gastric fermentation). In the non-ruminant herbivore, food is fermented in the large intestine (post-gastric fermentation).

of coprophagy cannot continue forever! In fact the rabbit eats only the soft or night pellets, which are formed when the food has passed through the gut once, and the hard or day pellets when formed from the soft ones are voided outside the burrow.

Why don't all animals digest cellulose using micro-organisms?

Ruminants use balanced food very inefficiently when compared with non-ruminants such as a pig or a bird; they convert less of the energy and protein in foods into their own tissue (Table 3.3). So to generate a given amount of animal tissue, they need more food than do omnivores; as with symbionts of legumes, the extra chemical steps cause lower efficiency.

The non-ruminant has, like humans, a requirement for 'essential' amino acids. It therefore needs high-quality proteins. It also needs starch rather than cellulose-rich foods, because it has no way of digesting this part of the plant. So we must consider not only the efficiency with which food is used, but also the quality of the food which has to be or can be supplied. The dilemma facing the farmer (and indeed the country) who wishes to provide humans with high-quality animal protein is as follows: the ruminant can make inefficient use of low-quality, energy-rich foods (cellulose) and proteins which are inexpensive, whereas the non-ruminant requires expensive high-quality energy and proteins in foods, but uses these more efficiently.

Table 3.3 The percentage efficiency with which various animals convert energy and protein foods to our food (based on various sources).

Animal	Food product	Protein	Energy
poultry	eggs	26	17
cattle	milk	24	16
poultry	meat	23	11
pigs	meat	14	14
cattle	meat	4	3
sheep	meat	4	2

Summary

The two major components of our diet, carbohydrate and protein, can each be manufactured in different ways.

Carbohydrates are produced by photosynthesis when carbon dioxide is fixed and reduced. When considering carbohydrate production, we must understand the merits of the two related ways in which plants do this. Alternatives facing us are these:

a we can choose the C_4 plant for tropical regions;

b we can choose the C_3 plant for temperate regions.

Proteins derive their nitrogen from a fixed form of the element. There are two sources, each with its own energy requirements, and hence two alternatives for us:

a we can use fixed nitrogen from chemical industries;

b we can use the symbiotic relationship between green plants and micro-organisms.

Animal protein provides further alternative patterns of food production. Animals must assimilate and utilise amino acids, but they can be given cheap plant proteins or inorganic nitrogen to eat, as well as expensive animal proteins (such as fish meal).

4 Domestication – things to date

Knowledge of the history of our food-producing organisms helps us in several ways. In general, the better we understand the biology of our present food producers, the greater are our opportunities to manipulate them to our advantage. History provides us with the pedigree for the organism which we use as a starting point for new breeding programmes. And history tells us of methods which led to the isolation of domesticated strains and breeds: methods which are of value when we plan a breeding programme.

Prehistory

Our early idea of the origin of plants and animals grown for food was that they were the gifts of the gods, and this notion has been slow to fade. We can now speculate as to when certain plants and animals were first domesticated by our ancestors.

Archaeologists have found plant and animal remains near early human dwellings; these have been subjected to procedures such as microscopy and general observation. Seed and bone remains can be compared with present-day domestic and wild populations, but this does not tell us when humans used these plants and animals.

The technique of **carbon dating** gives fairly precise estimates of how long ago plant or animal remains were made, and hence when the organisms were alive. A proportion of the carbon in the atmosphere (and thus in the carbon dioxide) is radioactive carbon-14 (^{14}C), so plants, and animals which eat them, incorporate ^{14}C. The ratio of radioactive ^{14}C to non-radioactive ^{12}C in the organic structures is defined while the organisms are alive; ^{14}C decays with time, so the ratio of ^{14}C to ^{12}C gradually decreases. Because we know the rate at which the decay takes place, we can use the ratio observed in the remains to estimate the age of these archaeological finds.

The closeness with which chromosomes and other chemicals in cultivated plants and animals correspond to their equivalents in present-day wild populations also provides important evidence concerning the origins of our cultivated plants and animals. We will return to this in more detail when we deal with the domestication of wheat.

What do we mean by 'domestication'?

Plants used by humans are said to be 'cultivated'; animals which feed and reproduce in close proximity to humans are 'domesticated'. We shall use 'domestication' to mean the process by which plants and animals become grown and used in our service, and not to mean 'taming'. Scientific use of the word 'domestication' has grown with our knowledge of the process; it was used formerly in the context of 'acclimatisation' and 'naturalisation' which refer to an ability to survive and breed in the wild. Domestic organisms can be protected from environmental rigours and may breed under our control, so the principles which underlie domestication have led to increased confidence in our use of the word.

The process of domestication

Darwin used domestication in plants and animals to support his theory of evolution (*Plants and Animals under Domestication*, 1896). He argued that what happened to these in the hands of humans may be a model for the origin of species. Ancestral crop types were very low 'yielding', and he suggested that these had been changed by humans using the 'better' members of the wild population: thus humans made selections. However, the Lamarkian controversy concerned domestication as well as natural speciation, and even Darwin wrongly suggested that the variation necessary for this selection was created by the environment.

It was Mendel, also using cultivated organisms, who published findings which led to an explanation of this variation, and who proposed certain precise laws stating the way characteristics are inherited. Mendel's work forms the basis of modern genetics, and genetics remains the cornerstone of plant and animal breeding. We shall return to it in Chapter 7.

How it all began

Domestication, which has greatly affected human development, has been governed largely by the unconscious activities of humans; only relatively recently has progress been made by deliberate interference. Humans began to grow food near settled dwelling areas instead of gathering it from the wild. They started by growing plants deliberately and were able thereby to extend their natural range of habitat; humans, plants and ideas spread, and new plants were brought into cultivation in new areas.

It is thought that there were two **nuclear zones** where agriculture started, one in the Old World, the other in the New. The first, in the region of the Euphrates and Tigris, emerged about 8000 BC and is known as the Fertile Crescent. The second, by contrast, started in 5000 BC in the area between New Mexico, Guatemala and Ecuador. As with other theories in biology, opinions on this differ; certain authors argue that some crops entered domestication in several regions at the same time.

The ideas and plants of cultivation spread from these regions where humans first abandoned nomadic life. Settlements where the early crops were grown provide archaeologists with a spreading track to follow; it leads from the Fertile Crescent to the west, south and east. Figure 4.1 shows just how gradual the movement was. Changes in the actual crops suggest that it was the ideas and know-how of cultivation which were most important.

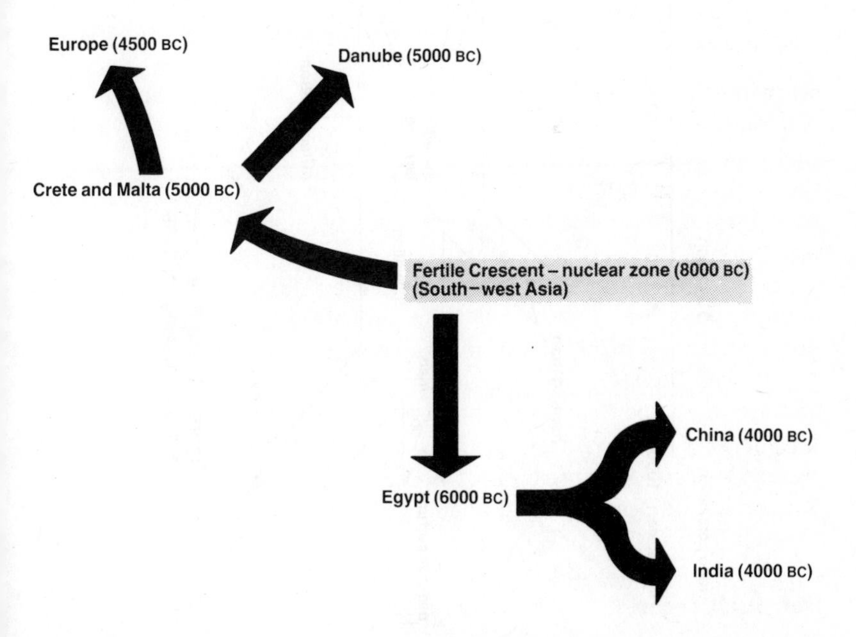

Figure 4.1 The directions in which plants and ideas moved from the Fertile Crescent.

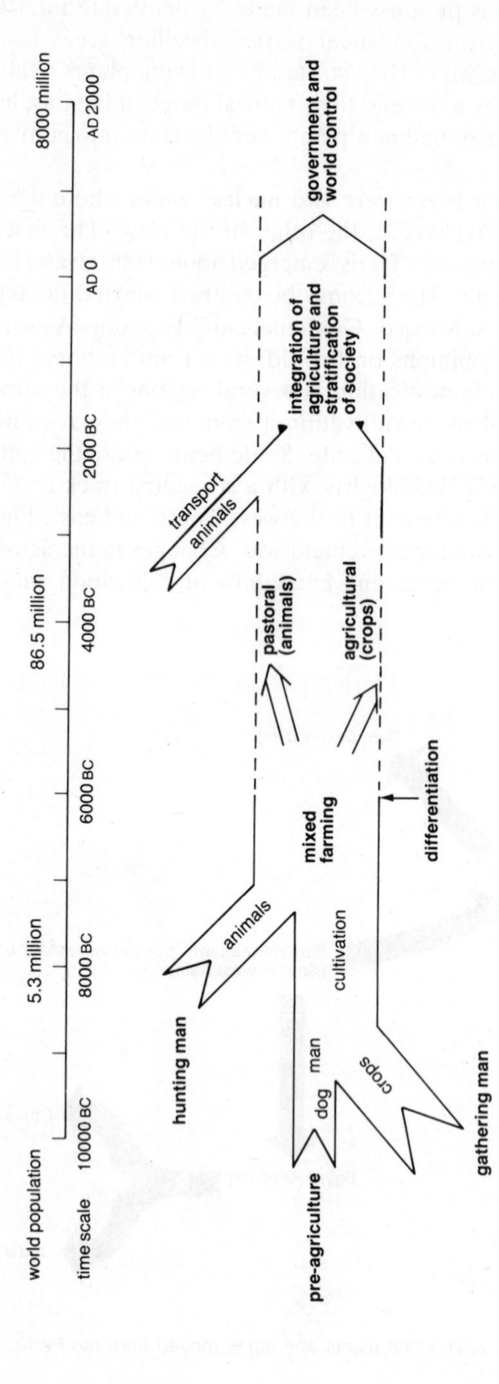

Figure 4.2 The dates at which grains from the Fertile Crescent reached certain parts of the world. (Based on Darlington, 1969.)

Regions of diversity

The crops and ideas of cultivation spread into parts of the world where small areas provided a large range of habitat. Near the equator, for example, the valley bottom contrasted greatly with the mountain side. Often known as 'regions of diversity', it is from these ten or so areas that some biologists claim to trace the development of each present-day crop (Table 4.1). 'Regions of development' might be more apt because the changes which took place led to many varieties; indeed, plant strains are still developing there and the variation in some of these populations is enormous.

Some organisations are trying to save some of this variation for the future use of plant breeders. The Food and Agriculture Organisation (FAO) is sampling these areas and conserving the variation in **gene banks**, storage facilities for live material. Seeds are often stored in controlled environments so that they remain viable, and living plant populations are maintained. From these populations, rich in variation, it is hoped to provide plant breeders with the 'wide genetic base' which is needed for today's breeding programmes.

Co-existence

So far, we have discussed the origins of domestication in plants only; as Figure 4.2 shows, this took place before that of animals, since humans could control them more readily. One notable exception was dogs, which associated with humans. As early as 15 000 BC, packs of wild dogs followed hunters and relied for survival on scraps of their food. They were accepted because they warned and protected humans against other aggressors. Humans exploited their natural characteristics and the relationship then became symbiotic; humans shared their prey when packs of hunting dogs attacked herds of herbivores. We now exploit dogs in hunting, retrieving, guarding, herding sheep and cattle and in the unnatural service of hauling loads.

Much later (around 6000 BC), the pig entered domestication by this same route of scavenging. While the pig ate scraps of food, it offered its flesh and hide to humans, and the limited services of clearing vegetation and hunting truffles, uses suited to the needs of established people rather than those of the earlier hunters.

Crops, such as tomatoes, marrows and some cereals, are thought to have originated from weeds growing around early human dwellings – so-called **habitation weeds**. Weeds grew on dung heaps, near dwellings and on burial grounds, favouring the level of nutrition provided by decaying waste.

Other weeds grew in crops – **crop weeds** – the weed becoming the

Table 4.1 Areas of domestication of major crops of the world. These regions of domestication are geographically both large and vaguely defined. There is no certain agreement between botanists who have studied and extended Vavilov's work. It seems that for each crop there may have been local regions of the source of wild ancestors, areas of domestication and areas showing great diversity, and that these regions may not have been the same area
(based on Darlington and Simmonds).

1 South-west Asia – Near East	
† Bread wheat	*Triticum aestivum*
† Barley	*Hordeum vulgare*
Lentil	*Lens culinaris*
Rape	*Brassica campestris*
Opium poppy	*Papaver somniferum*
Melon	*Cucumis melo*
Carrot	*Daucus carota*
Fig	*Ficus carica*
Pomegranate	*Punica granatum*
Cherry	*Prunus avium*
† Grape vine	*Vitis vinifera*
Date palm	*Phoenix dactylifera*
† Pea	*Pisum sativum*
2 Mediterranean	
Broad bean	*Vicia faba*
† Cabbage etc.	*Brassica oleracea*
Swede	*Brassica napus*
Olive	*Olea europaea*
† Sugar beet	*Beta vulgaris*
Hop	*Humulus lupulus*
2a Europe	
† Oats	*Avena sativa*
† Rye	*Secale cereale*
Currants	*Ribes* spp.
Raspberries	*Rubus* spp.
3 Ethiopia	
† Finger millet	*Eleusine coracana*
Coffee	*Coffea arabica*
3a Central Africa	
† Sorghum	*Sorghum bicolor*
Oil palm	*Elaeis guineensis*
4 Central Asia – Afghanistan	
Common millet	*Panicum miliaceum*
Buck wheat	*Fagopyrum esculentum*
Hemp	*Cannabis sativa*
Alfalfa	*Medicago sativa*
Pear	*Pyrus communis*
† Apple	*Malus pumila*

5 Indo–Burma	
† Rice	*Oryza sativa*
Eggplant (aubergine)	*Solanum melongena*
Cucumber	*Cucumis sativus*
† Mango	*Mangifera indica*
Pepper	*Piper nigrum*
6 South-east Asia	
† Yam	*Dioscorea* spp. (other areas in Africa and South America)
† Banana	*Musa* spp.
† Coconut	*Cocos nucifera*
† Citrus fruits	*Citrus* spp.
† Sugar cane	*Saccharum* spp.
7 China	
† Soya bean	*Glycine max*
Apricot	*Prunus armeniaca*
Peach	*Prunus persica*
Tea	*Camellia sinensis*
8 Mexico	
† Maize	*Zea mays*
Kidney bean	*Phaseolus vulgaris*
Red pepper	*Capsicum annuum*
† Upland cotton	*Gossypium hirsutum*
Sisal	*Agave sisalana*
8a USA	
† Sunflower	*Helianthus annus*
8b Central America	
† Squash, gourds etc.	*Cucurbita* spp.
9 Peru	
† Sweet potato	*Ipomoea batatus*
† Potato	*Solanum tuberosum*
† Tomato	*Lycopersicon esculentum*
Tobacco	*Nicotiana tabacum*
Lima bean	*Phaseolus lunatus*
Quinine	*Cinchona* spp.
9a Brazil – Paraguay	
† Tapioca – Cassava	*Manihot esculenta*
† Groundnut	*Arachis hypogaea*
Cacao	*Theobroma cacao*
Pineapple	*Ananas comosus*
Brazil nut	*Bertholletia excelsa*
Rubber	*Hevea brasiliensis*

(† indicates annual harvest greater than 100 million tonnes) crop once humans recognised its food value. Rye probably developed in this way; it originally contaminated wheat crops in the Mediterranean region, and when contaminated wheat seed was grown in Russia, conditions favoured rye rather than wheat. Even today, in certain mountain regions, the proportion of rye in wheat crops increases as one climbs the mountain. Barley also is thought to have been a crop weed, in primitive wheat (Emmer wheat).

Humans' early selections

At first, humans domesticated plants and animals which they could control – grains which were trapped by their method of harvest, animals which they could restrain. Domestication depended on the organism conforming to the human pattern of farming. Only later did humans select on the basis of high yield.

Wild grains of barley, wheat, oats and rye were well adapted for survival and dispersal. The rate at which they germinated was variable, so that the plant population could overcome any setback occurring while plants were small and delicate. Grains ripened unevenly, which made best use of the dispersing agents; chaff often had long, barbed projections which tangled in the fur of passing animals, and the grain-containing ears fragmented, thus allowing them to disperse seeds in small numbers. All such adaptations equipped plants for survival in the wild.

These attributes were of little value to farmers – in fact they did not collect plants which had them. Late germination would mean late harvest; farmers would harvest just once so those plants which were still flowering would be wasted. Similarly, plants which took a long time to mature would not contribute to the yield. If the ears fragmented, seed was lost. This led Darlington to suggest that three pressures operated upon crops during domestication; sowing conditions, tillage conditions (ploughed, manured land caused early vigorous growth) and harvesting conditions.

The evolution of wheat

Present-day wheat varieties differ greatly from those first domesticated, which were stunted and low yielding. However, it is here that we can start piecing together the evolution of modern wheat. Farmers near the Fertile Crescent grew mixtures of different types of wheat, including Emmer and Einkorn. Emmer wheat was probably carried northwards towards Persia as ideas and crops left the Fertile Crescent, then bred

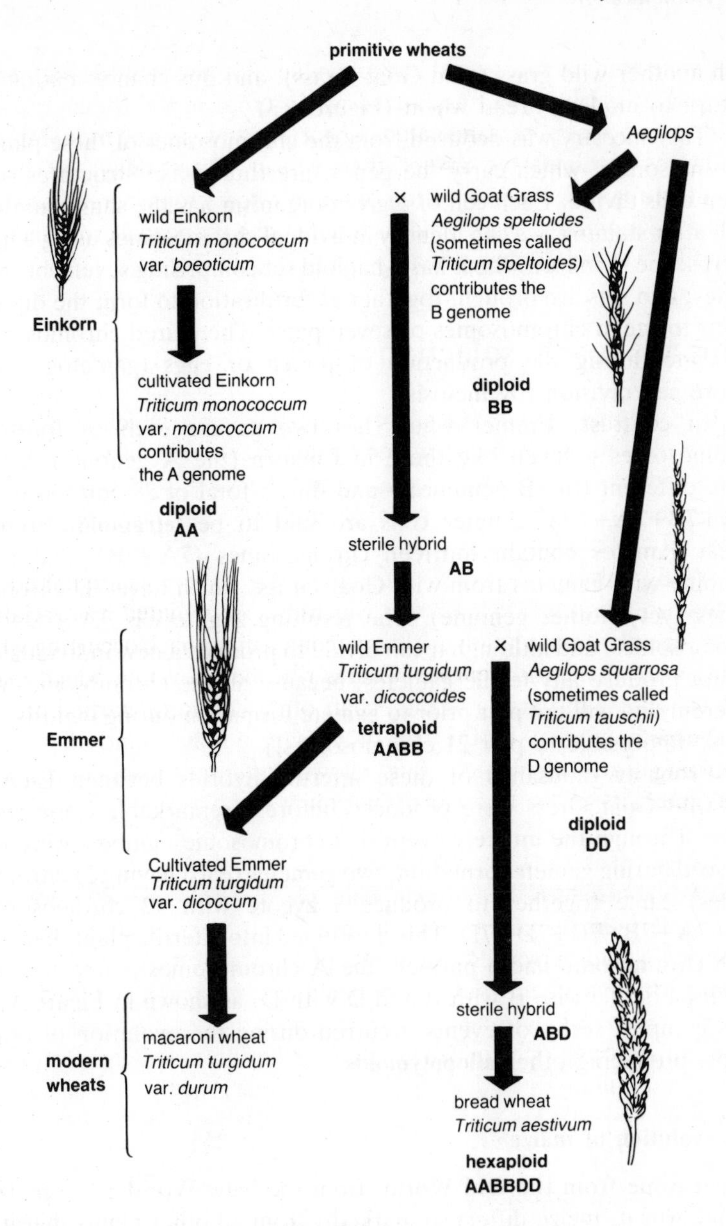

Figure 4.3 The evolution of wheat (*Triticum*).

with another wild grass (wild Goat Grass), and this chance mating led in turn to modern bread wheat (Figure 4.3).

This ancestry was deduced from the chromosomes of these plants. Chromosomes, which carry the genes, are thread-like structures seen when cells divide. Each cell of a given organism has the same number, and, after staining, we can identify individual chromosomes using a light microscope. Einkorn wheat has a haploid set comprising seven chromosomes; two sets are brought together at fertilisation to form the diploid set of fourteen chromosomes or seven pairs. The paired chromosomes associate during the production of pollen or eggs (gametogenesis) before cell division (by meiosis).

In contrast, Emmer wheat has two groups each of fourteen chromosomes – seven like those in Einkorn (the A genome), seven quite different (the B genome) – and thus a total of 28 chromosomes (7A+7B+7A+7B). Emmer cells are said to be **tetraploid**. Emmer wheat gametes contain fourteen chromosomes (7A+7B), and may combine with gametes from wild Goat Grass, which have 7D chromosomes (yet another genome). The resulting zygote has 7A+7B+7D chromosomes, and although it can divide to produce a new individual, it cannot produce any fertile gametes, because all the chromosomes are different and will not pair prior to gamete formation during meiosis. (It is also impossible to pair 21 chromosomes!)

Probably thousands of these infertile hybrids between Emmer and wild Goat Grass were produced before a remarkable event took place. Through the unlikely event that chromosome numbers were not reduced during gamete formation, two gametes (each having 21 chromosomes) came together to produce a zygote with 42 chromosomes 7A+7A+7B+7B+7D+7D. This developed into a fertile plant, because each chromosome had a partner (the A chromosomes paired to form seven pairs, likewise B with B and D with D, as shown in Figure 4.3). This complex series of events occurred during the evolution of other crops, producing other **allopolyploids**.

The evolution of maize

Wheat came from the Old World; from the New World came maize. Unlike wheat, maize differs so markedly from all other plants that it is classified as a separate genus, *Zea*. Either similar wild ancestors became extinct or maize changed radically; theories of the evolution of maize are provided with few clues. It is most likely that maize derives from an annual called teosinte (*Euchlaena mexicana*). This has a brittle stalk to the ear and the grains are small and enclosed in small leafy husks; they are said to be ‘hulled’. In contrast to teosinte, the maize cob has a robust stalk with the whole enclosed in large husks. Evidence such as the fact that common forms of teosinte hybridise with maize and that both plants have ten chromosomes suggests a link between the two plants.

Domestication of herbivores

Humans domesticated plants first; it was much later that they stopped hunting and incorporated grazing animals into their agricultural system (Figure 4.2). Ancestors of today's herbivores initially forced themselves on humans. Having invaded their crops, they provided a ready alternative when natural food was scarce; domestication then followed the pattern set by plants. No doubt humans provided them with restricted grazing and fenced in these pastures. Herbivores were naturally placid, and because of their bulky diet, spent much of their time grazing. They were ideal candidates for domestication.

In the early stages, humans chose inevitably those individuals which were easily managed; high-spirited and aggressive animals escaped or were chased away. Humans, in their domestication, reversed some of the evolutionary pressures imposed upon wild populations where the spirited individuals survived. Later on, humans kept and bred from the productive rather than the placid.

The criteria for fitness changed; humans became interested in the provision of meat and later milk, wool and a means of haulage. Cattle, sheep and goats which released milk were favoured, as were sheep which retained their wool so that they could be sheared before it was shed naturally. Cattle which were docile enough to be harnessed to the plough were also selected and bred from. Domestication was under way; several animal species each exploited a slightly different environment in the human agricultural system and *each* was capable of providing humans with most, if not all, of their meat, milk, haulage, skin and wool (Table 4.2).

Changing agricultural patterns

Between 6000 and 3000 BC, both humans and their agricultural systems became more specialised. Two basic patterns emerged: one (in Central Asia, Arabia, Sudan and Southern Africa) was animal-dominated, the other (in Europe, Egypt, India, China and America) was plant-dominated. People engaged in these branches of agriculture acquired specific skills which changed continually to gain the maximum yield from the land (Figure 4.2).

Recent animal domestication

Some animals were domesticated only recently, the result of a deliberate act by humans to provide themselves with specialised transport to increase their mobility. Ass, horse, camel and elephant were deliberately domesticated around 3000–2000 BC. As human interests became more extensive, humans needed to plough fields, to transport people and goods, and some to fight wars so that they could exploit more distant regions or defend their own.

Specialised mobility often required the repeated training of wild animals, a process still carried on in southern Asia with the elephant. Elephants are and were more often captured from the wild than born into captivity. The training of these animals for work is really taming rather than domestication.

Table 4.2 The times and places of animal domestication (based on Darlington, 1969).

Animal	Time of domestication (years BC)	Place of domestication
dog	15 000	many places
reindeer	10 000	many places
goat	7 000	Persia
sheep	6 500	Caspian steppes
cattle	6 000 2 500	Anatolia Indus
pig	6 000 2 000	Anatolia China
ass	4 000	Egypt
elephant	2 500 280	Indus Egypt
horse	2 250	Caspian steppes
water buffalo	2 000 1 000	Indus China
camel (dromedary)	1 200	Arabia
camel (bactrian)	500	Central Asia
yak	1 000	Nepal

Although these later domestications arose from the need for transport, humans did not ignore other qualities of these animals, which could now contribute to meat, hide and milk production.

The control of breeding

During the process of domestication there was migration of plants, animals and ideas, and the cultivated populations often bred with their wild relatives. This contributed to the vast quantity of variation shown in the 'regions of diversity', variation which provided the opportunity for selection. The mating patterns of plants and animals often changed during the process.

Humans gradually gained control of the mating of animals as these conformed more and more to human patterns of farming; with plants this was not so easy, though changes did take place. Control of pollination presents problems even to the modern plant breeder. When a new environment did not favour the pollinating insects, a plant relied for pollination on self-fertilisation. The tomato provides a good example of this: in its native regions it was cross-pollinated; in Europe, however, it will only self-pollinate. The cross-pollination found among the ancestors of wheat also has been replaced in modern bread wheats by self-pollination.

With the change from cross-pollination to self-pollination, the amount of genetic variation has been reduced; the plant has become an inbreeder (see Chapter 8). This reduction of variation is desirable to farmers; it means that all the plants are uniform and therefore show the same response to fertiliser treatments and harvesting methods. But it gives problems to the plant breeder, as we shall see; he must break down self-fertilisation at the start of a breeding programme so as to achieve an initial mating between two potential parents.

Modern cultivated strawberries (*Fragaria ananassa*) are thought to have resulted from the reverse process. When they were both grown in Europe, the breeding barriers between ancestral species of strawberry broke down. The South American plant (*Fragaria chiloensis*) and the North American plant (*Fragaria virginiana*) produced their first hybrid in Europe in 1750 and this gave rise to our modern plants.

Plants sometimes change their breeding in response to environmental factors such as day length. Flowering, in some plants, is initiated by a particular proportion of daylight in the 24-hour cycle. Our present varieties of potatoes have developed from South American plants which produced potatoes (tubers) under short day-length conditions. Main crop varieties, grown for high yield and harvested in early autumn, retain this response. 'Early' varieties are harvested in the long days of early or mid-summer; by deliberately selecting early-maturing plants, we have completely changed the plant's original response to day length.

Some crops have stopped reproducing sexually (so slowing evolution) as they do not use gametes in seed-formation, a state of affairs known as **apomixis**. It resembles sexual reproduction in that flowers, egg cells and pollen cells are produced, as are fertile fruits, but the seed and embryo develop from a cell which is not the zygote. This rather peculiar method of reproduction is found in oranges, other citrous fruits and sugar cane.

Breeding from the best

Humans soon realised that their fortunes depended on the productivity of their plants and animals. Their skills in husbandry governed their own wealth and that of their children. One skill which came to the fore was the taking and breeding from the best organisms in captivity. Thus the unconscious selection in the first domestication of plants and animals gave way to what Darwin called 'methodical selection'. Much of the early selection was haphazard; there was little idea of the qualities which contributed to yield. An important result of these early selections was uniformity; hence the many breeds and varieties we see today. Plant and animal breeding, although in its infancy, had been born.

Summary

The origins of many of our domesticated plants and animals can be traced to living or long-dead species; this knowledge benefits us in several ways:

a living relatives of cultivated species may be used in present breeding programmes so as to provide a greater base of genetic variation;

b new species of crop plant may be synthesised;

c the conditions for reproduction of domesticated species may be established by reference to that of related species;

d we may gain an understanding of the way that organisms which benefit us have in turn affected our own development from prehistory.

5 Optimum use of our biological resources

The process of domestication has provided us with a wide range of species, breeds and varieties which make food of a particular quality under certain conditions. To improve production from these organisms, two complementary techniques are available – increasing the availability of building blocks for growth (changing the environment to ease the flow of nourishment to the organism) and increasing the genetic merit of the organism (to convert available building blocks into body tissue more efficiently).

Bottlenecks in the flow of building blocks into the organisms are governed by the law of limiting factors. To mitigate its effects, we must improve the environment and thereby ensure that shortages of one factor do not cause the waste of valuable nutrients.

This is only one side of the story. Alterations in the feeding environment may be expensive and farmers may choose not to make them: they may get less money for the increased yield than they spent in improving the availability of the nourishment (a point that we shall return to in Chapter 9).

The law of limiting factors

If one factor is known to be limiting growth, and supplies of this factor are increased in uniform steps, there will be less and less improvement in growth per step, because the availability of *other* substances becomes increasingly important. Mitscherlich, a German scientist, formulated this principle after work on a fertiliser treatment (a situation which we shall consider in this chapter). Figure 5.1 depicts this conclusion of the law of limiting factors, known as the **law of diminishing returns**.

Nutrients in the feeding environment may be considered to be

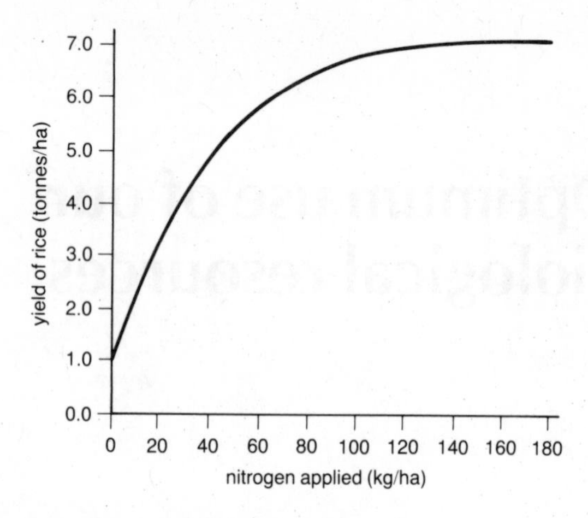

Figure 5.1 The effect on rice of additions of nitrogen to the growth medium. Rice varieties used were short, stiff-strawed and non-lodging.
(From the *Rice Production Manual.*)

factors limiting growth, so production can be improved either by adding nutrients directly or by removing competitors for nutrients. Technical methods are available to the grower, ranging from soil and animal food analysis to selective chemical weedkillers and drugs. Farmers must have great sympathy with their plants and animals; their skills are based upon scientific fact, but are tempered by experience. Their role is best described by the little-used word 'husbandry', which means 'careful management'. This management starts with the soil.

The environment of the soil

Water in the soil

Of the water taken up by a plant, very little is used in photosynthesis; most is lost through the stomata. Stomata allow carbon dioxide into the leaf for photosynthesis but cannot distinguish between admission of carbon dioxide and the loss of water. When water is in short supply and the stomata are closed to reduce water loss, the limited uptake of carbon dioxide restricts growth.

At the other extreme, when all the small pores between the soil particles fill with water, the soil is said to be saturated. With prolonged saturation, roots suffocate; the soil becomes anaerobic because oxygen diffuses slowly. Waterlogging for 24 hours may reduce the oxygen content of the soil from the normal 18–20 per cent level to less than two per cent. Alcohol formed by anaerobic respiration progressively kills

Figure 5.2 The effect on the growth and yield of peas (*Pisum sativum*) of a short period of waterlogging.
(Courtesy of Agricultural Research Council Letcombe Laboratory.)

Table 5.1 The effect on the growth and yield of peas (*Pisum sativum*) of a short period of waterlogging.

	Control in freely draining soil C	Growth stage at which plants were waterlogged for a five-day period			
		T1 Vegetative (3–4 leaves)	**T2 Flower-bed (6–7 leaves)**	**T3 Flowering (9–10 leaves)**	**T4 Pod-filling (9–10 leaves)**
Seed dry weight (g/plant)	9.1	4.3	0.7	2.3	2.7
Total dry weight (g/plant)	18.1	7.4	2.0	6.1	7.4

the roots, and so the yield of ill-adapted crop plants grown in low-lying fields is reduced (Figure 5.2 and Table 5.1).

Generally, water drains from soil until the gravitational forces of drainage are balanced by an expression of surface tension, as capillary action, which holds water around soil particles. The water content of the soil at the point of balance is called **field capacity**. Plants can readily remove the water they need and gaseous oxygen may freely move within the soil atmosphere. If the water content is not replenished, the plant experiences more and more difficulty in removing water from the soil because the forces of surface tension increase. Although the forces exerted by transpiration are very powerful, a time is reached when the plant wilts. This is not uncommon on warm, sunny days; the plant wilts during the day but recovers during the night when the need for water is reduced. Eventually there may be so little water available that the plant may not revive, even when enclosed in a darkened chamber (which causes the stomata to close and transpiration to stop). Under these conditions the soil is said to be at **permanent wilting point**; soil particles hold the remaining water so tenaciously that no water is accessible to the plant.

The working range of 'available water' in the soil is between field capacity and permanent wilting point; this range varies from soil to soil (Figure 5.3). At field capacity, the water content of a clay soil is much

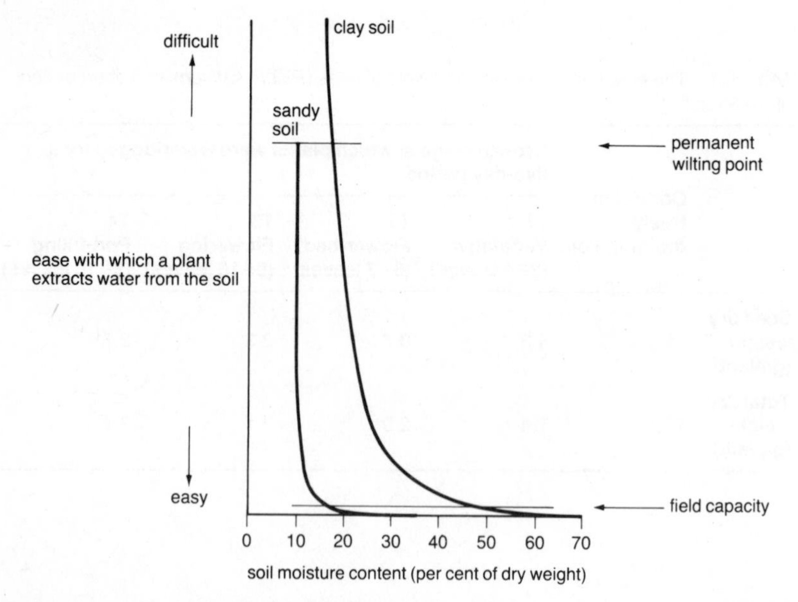

Figure 5.3 Water-retention properties of soils.

greater than that of a sandy one, partly because the particles are smaller (so more water is held by capillary action), partly because these small particles have a fine lattice structure which itself holds water. Clay soil allows more of that water present at field capacity to go to the plant.

There are several methods of increasing the availability of water to plants. Perhaps the most obvious is to irrigate the land; this has dramatic effects on the productivity of arid areas, but can also improve yields in temperate regions. It requires a large and reliable water supply.

Traditionally, organic matter is added to sandy soil to increase the water reserve. The sponge-like nature of the organic remains enables the soil to hold water; and the humic acid, formed when these remains decay, coats soil particles as a water-retentive film. Because organic matter decays continually, we must regularly add more suitable material.

Water is lost from the soil not only via plants but also simply by evaporation from the surface. This can be reduced by placing an artificial barrier at the surface. One such scheme is to allow plants to grow through holes in a sheet of black polythene (Figure 5.4); rain water enters the soil via the holes and may leave only via the plant. Black polythene has the added advantage that it warms up the soil. Coarse organic matter (such as peat or straw), sand or polythene scrap can provide a barrier and reduce evaporation. These methods are known as

Figure 5.4 Use of black polythene as a mulch (J.B. Land).

Table 5.2 Quantity of minerals removed by a crop in one year (United Kingdom).

Mineral	Quantity removed (kg/ha)	Mineral	Quantity removed (mg/ha)
nitrogen (NH_4^+, NO_3^-)	100	iron (Fe^{2+})	600
potassium (K^+)	100	manganese (Mn^{2+})	600
calcium (Ca^{2+})	50	zinc (Zn^{2+})	200
phosphorus (PO_4^{3-})	15	boron (BO_3^{3-}	200
magnesium (Mg^{2+})	15	copper (Cu^{2+})	100
sulphur (SO_4^{2-})	30	molybdenum (MoO_4^{2-})	10
		cobalt (Co^{2+})	1

mulching. The simplest and most widely used technique of forming a barrier is to **cultivate** the soil surface.

Minerals in crops

Crops remove a substantial quantity of minerals from the soil; Table 5.2 shows how much.

If plant tissue is burned, just over twenty minerals will be found in the ash. There is continual research to establish whether these are all essential for growth. Sodium and chlorine should now be added to those in Table 5.2; these are now considered to be essential.

The need for all of these elements can be shown by water culture or **hydroponics**, first used by John Woodward in the seventeenth century. It entails growing the plant in water rather than soil (see page 181); minerals known to be essential for growth are added to the water in the correct proportions and quantities, with the exception of the mineral under investigation. The characteristic effects which develop in each case are called **deficiency symptoms**. They are well documented; the experienced farmer or advisor can diagnose deficiencies simply by visual inspection (Figure 5.5).

The major technical problem with this method of investigation is the impurity of chemicals and of water. As the methods of purification have improved, so the list of essential minerals has lengthened, and with it our knowledge of the abnormal biochemical reactions taking place in such plants.

Availability of minerals to the plant

By the time a crop shows deficiency symptoms, yield has already been drastically reduced; the plants are usually about to die. So for maximum production, it is the *possibility* of deficiency which must be avoided; we

Figure 5.5 Iron deficiency: the leaf on the left is healthy; the centre leaf shows iron deficiency; and the leaf on the right shows old age (senescence). Whereas a senescent leaf becomes yellow all over, a leaf deficient in iron is yellow only in the areas between the veins, the veins themselves remaining green.
(Courtesy of B. J. W. Heath.)

must know the availability of each mineral to the crop, and so we must understand nutrient cycles.

Minerals generally exist in at least four forms in the soil. One is in **organic plant and animal remains and excreta**; these may store and slowly release minerals (such as SO_4^{2-}, PO_4^{3-} or NO_3^- ions), while at the same time the living plant may be storing and gradually removing minerals (as we saw in Table 5.2).

The second form is the **soluble fraction**, minerals in soil solution; these are taken up by the root hairs from the soil water. They are the *available* minerals.

Thirdly, there is a large reserve of minerals in the soil which are *not* immediately available to the plant; they are in ionic form and in equilibrium with the soluble fraction, and form the **exchangeable fraction**. We can estimate this fraction experimentally by saturating the soil with one ion species, which literally forces all other similarly charged ions off the charged sites in the soil, and the removed ions are collected as an exchangeable fraction. By contrast, the **non-exchangeable reserve** of ions *cannot* be removed by ion competition (see Figure 5.6).

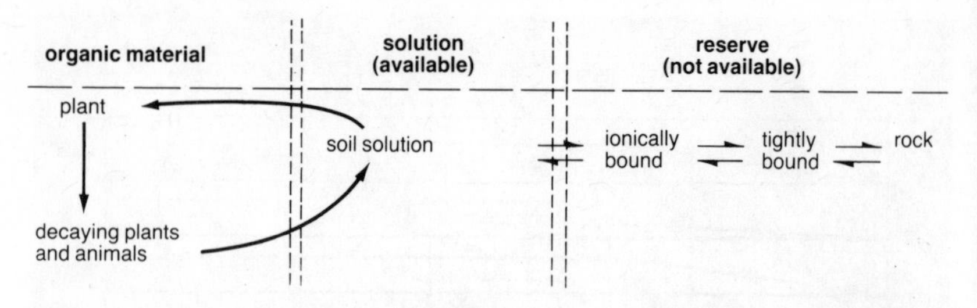

Figure 5.6 The relationship between the forms of minerals present in soils.

For a mineral such as potassium, only a small proportion of the total soil content is in solution; the rest is held in reserve forms. On the other hand nearly all of the nitrogen is in solution; NO_3^- and NH_4^+ ions are bonded only lightly to the soil particles, and no rock contains quantities of nitrogen accessible for plant growth.

When minerals are applied to a soil as a fertiliser, only a portion will be recovered in the crop – the rest is either locked up in the mineral cycle or lost. For example, up to 80 per cent of the PO_4^{3-} ions suitable for plant uptake become insoluble immediately on contact with the soil, while only 30 per cent can be recovered over several years in grass crops. It is therefore argued that phosphatic and potassium fertilisers are best applied each year. Large, less frequent applications result in poor growth at the end of the period. Nitrogenous fertilisers, though, should be applied at the start of a growing season in temperate regions for a different reason, because NO_3^- and NH_4^+ ions are held only weakly, and are easily lost in drainage water. As well as the loss to plants, this may cause pollution of water: a particular danger with NO_3^- ions which may be converted under anaerobic conditions to the very toxic NO_2^- ions.

Acidity of soil water

The pH of soil, be it low (acidic) or high affects the physical properties of the soil, such as drainage, growth of micro-organisms and availability of certain minerals. Optimum levels of minerals for plant growth are found in neutral or slightly acidic soils, so most crops are grown on such soils. Plants grown on acidic or alkaline soils generally show either mineral toxicity (excess of an ion) or deficiency (lack of an ion); the solubility – and thus availability – of most minerals is governed by the proportion of hydrogen (H^+) ions to hydroxyl (OH^-) ions. Figure 5.7 shows the effects on mineral solubility of varying pH.

Iron is a mineral which in alkaline soils becomes unavailable to plants (Figure 5.7); it is an example of the relationship between pH and

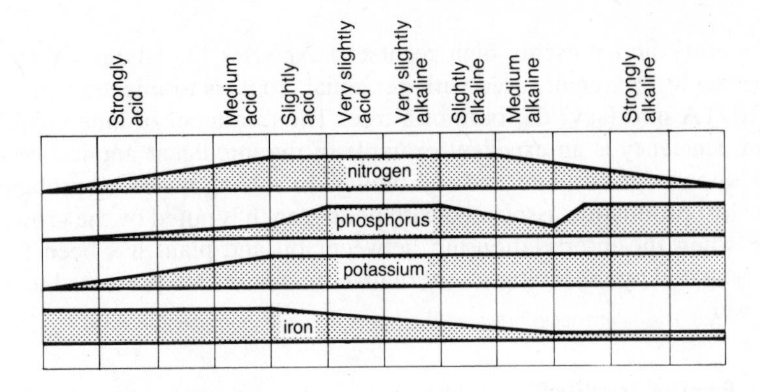

Figure 5.7 The solubility of ions over a range of pH. (Based on Sutcliffe and Baker.)

mineral availability. In the soil, iron is soluble under acidic conditions but is precipitated as a hydroxide in alkaline soil. It is needed in quite large quantities for plant growth, being part of the cytochrome molecules and essential to the production of chlorophyll. A deficiency of iron gives the characteristic symptoms of yellowing leaves; yellow patches develop between the veins (interveinal chlorosis).

Certain lime-hating (**calcifugal**) plants, such as rhododendrons, favour acidic soils because these provide high enough levels of iron in solution. Fortunately, few commercial annual crops are lime-haters; most tolerate an acceptable pH range. Their tolerance may have been a factor in their domestication. Problems do sometimes develop with perennial crops such as apples, as the roots of mature plants may penetrate a lime-based subsoil. Also, farmers sometimes use lime to create a more alkaline soil, because slightly alkaline material tends to be washed out of soil or removed by growing plants; lime ($Ca(OH)_2$), being alkaline, redresses this. So too much added lime, irrigation with alkaline water or natural reasons may result in low ion availability, and all cause 'lime-induced iron chlorosis'.

We can cure iron deficiency in several ways and thereby promote healthy growth. One method, the addition of iron sulphate, does not work well, because the iron is locked up again as the hydroxide. Regular and liberal applications will eventually reduce the pH by salt hydrolysis and so leave iron in solution. A better approach is to ensure that the iron remains soluble and available to the plant. This we can do by incorporating it into an organic complex such as EDTA which can be taken up by the plant. An alternative is to avoid contact with the soil, by applying an iron salt in a foliar spray. Foliar feeding requires quite small quantities of the mineral and is extremely effective, but its effect is only temporary and presents high recurrent expenses for labour. A more permanent treatment, used with perennial crops, is to inject a pellet of FeEDTA into (say) the bark of a tree. The treatment of lime-induced iron deficiency is an excellent example of the intelligent application of our knowledge of plant and soil biology in solving a serious problem. The lost production associated with poor growth is noted by the grower, and when the interrelationship between soil and plant has been fully investigated by the biologist, alternative treatments are presented to the grower.

The form of fertiliser

Plant nutrition can be applied to the soil in two basic forms, one of organic origin, the other of inorganic ions. Both can provide nitrogen, phosphorus and potassium. Organic fertilisers supply these in predictable small quantities, with organic matter and some trace elements; inorganic fertiliser can be tailor-made to requirements, with trace elements added as necessary.

Organic fertilisers are often called **manures**. Their use improves the water retention of sandy soil, and the improved crumb structure helps drainage and hence increases the availability of oxygen in a clay soil. Inorganic fertiliser has very little effect on soil structure.

Problems associated with the use of organic manures stem directly from their high organic content. Consider the fate of straw (3500 kg/ha) remaining from a crop of wheat. The farmer often burns this, and any organic matter which might improve the soil is thereby lost, as is nitrogen (17 kg/ha); by contrast, phosphorus (3 kg/ha) and potassium (30 kg/ha) are returned to the soil. The alternative – to incorporate straw into the soil – is difficult because of its bulk, and it locks up nitrogen (one part nitrogen for every twenty parts carbon) while micro-organisms break down the straw. High levels of organic matter alone actually reduce soil fertility in the short term, therefore; any improvement in plant growth comes from improved physical characteristics of the soil. Fortunately, many organic manures contain animal waste products which help to supply minerals, especially nitrogen, to the soil. We can now see why the farmer burns straw and yet uses organic manures; perhaps we should, as a society, investigate what use we might make of this energy-rich waste product (see Chapter 11).

Inorganic fertilisers can be formulated to meet the needs of the crop. These depend in part upon the quantities of ions in the soil and the fertiliser requirements of particular crop plants. Leafy crops such as cabbage and grass need nitrogen; root and tuber crops such as potatoes and carrots need phosphates to ensure adequate growth of storage organs; fruit growth and development need potassium. For the most part, it is the level of additional nitrogen which varies most with the crop grown; legumes, as we saw in Chapter 3, do not require any fertiliser nitrogen.

Organic and inorganic fertilisers should not be thought of as alternatives. They are applied to soil to optimise it as a medium for plant growth by balancing the availability of plant nutrients (including oxygen and water). The plant simply absorbs inorganic salts through its root hairs, without regard to their origin; these must be available in an environment suitable for plant growth.

Another important factor in deciding whether or not to use a fertiliser is the cost of applying it (Chapter 9).

Removal of competitors

In our efforts to ensure that energy and nourishment actually reach the food producers, another approach is to remove their rivals for these resources. A major loss of plant productivity is through competition by unwanted plants; we call any unwanted plant a weed! Equivalent animal competition is shown, for instance, by rabbits at pasture and by rats or other pests invading food stored by humans for themselves and their animals. Agriculturalists try to remove such competitors, be they plant or animal.

Weedkillers

Plant husbandmen must remove weeds from the soil in which the crop grows. Hand weeding, while possible, is very time-consuming and hence expensive, and may not prove very successful in the long term.

Selective weedkillers were in use as long ago as 1896, when copper sulphate solutions were employed to eliminate broad-leaved weeds from cereal crops in France. In the 1930s, sulphuric acid was used. Some 80 chemicals are now formulated into about 450 products marketed by the large chemical companies; these provide the farmer with a vast range of specific, total and selective weedkillers.

Types of weedkiller

It is not necessary to review this whole range of chemicals in order to highlight the biology of the basic modes of action. There are three types of weedkiller (Figure 5.8) which we will consider in turn.

The first sort is the **total contact herbicide** which kills all the green

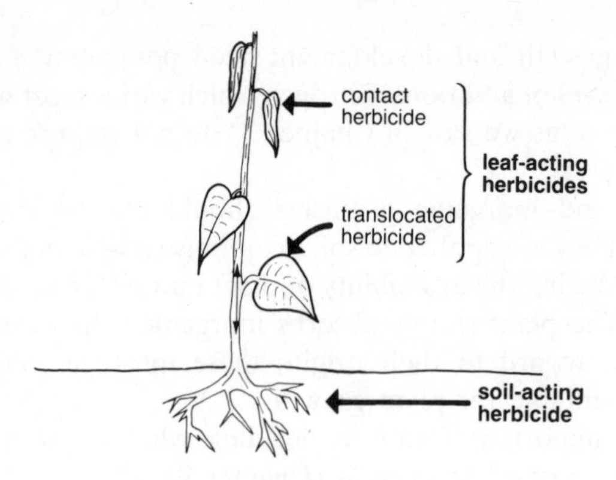

Figure 5.8 Types of weedkiller.

parts of all plants growing in the treated area. Such a chemical rapidly produces a very stable seed bed devoid of vegetation without cultivation, so avoiding erosion and allowing another crop to be planted quickly. With the weeds killed, there is no need for ploughing and further cultivation. Technology then solves the problem of 'sowing' the seed; the soil is cut open, the seed put in the groove and this is then rolled closed. A field grazed one day can be sown with seed the next. A prerequisite of such herbicides is that there is no lasting effect. Paraquat is an example.

Persistence of action becomes an advantage when a total herbicide is needed to kill *all* the vegetation on, say, a path or patch of gravel. Crude poisons can also be used; sodium chlorate exerts its effect in the soil for a year or two, so that *no* plants can be grown.

With the second group of weedkillers, **soil-acting herbicides**, a chemical is applied to the soil and left to act over a long period, continuing to act on weeds growing between crop plants; these herbicides must be selective. The selectivity is usually achieved by the chemical remaining in the top two centimetres or so of soil, where it kills germinating seeds but does not affect deeper-rooted plants. These chemicals are applied to the cleared soil around deep-rooted crops, which can thus be kept weed-free. Simazine is one such herbicide.

There is tremendous interest in the third group, **selective herbicides**. These can weed a growing crop without any of the damage associated with hand weeding. Selective herbicides are applied to most cereal crops grown under intensive conditions; the crop and its stage of development often determine the particular chemical used. The most widely known are those applied to lawns, including 2,4–dichlorophenoxyacetic acid (2,4-D), 4-chloro-2-methylphenoxyacetic acid (MCPA), and 2,4,5-trichlorophenoxyacetic acid (2,4,5-T), used in

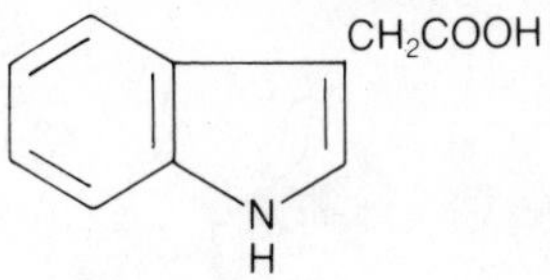

CH_2COOH

indole-3-acetic acid (IAA)

CH_2COOH

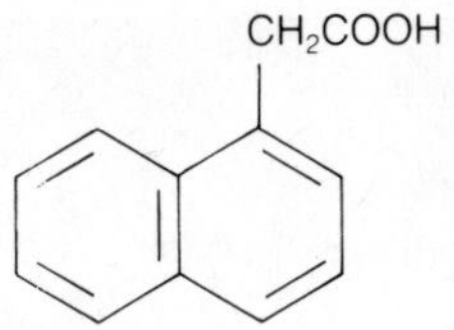

1-naphthaleneacetic acid (NAA)

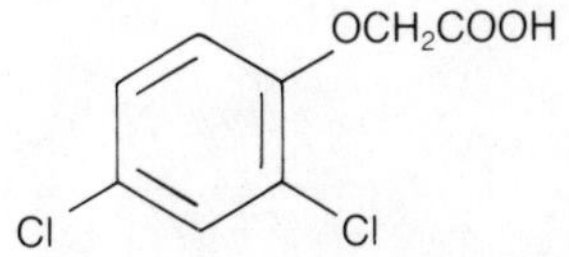

OCH_2COOH

Cl Cl

2, 4-dichlorophenoxyacetic acid (2, 4-D)

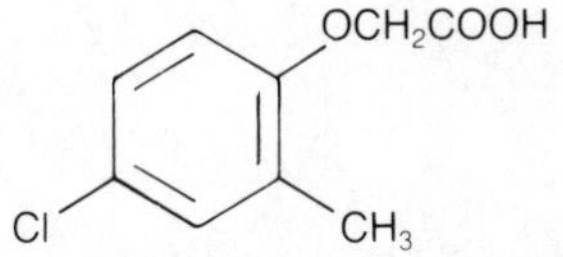

OCH_2COOH

Cl CH_3

4-chloro-2-methylphenoxyacetic acid (MCPA)

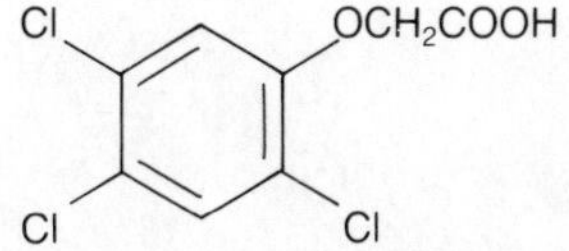

Cl OCH_2COOH

Cl Cl

2, 4, 5-trichlorophenoxyacetic acid (2, 4, 5-T)

Figure 5.9 The molecular structures of artificial and naturally occurring growth regulators.

vast quantities by the United States Air Force to defoliate large areas of forest in Vietnam. Figure 5.9 shows how similar these are to the naturally occurring plant hormone indole-3-acetic acid (IAA). Their action exploits the fact that they are taken up by broad-leaved plants, translocated and exert their effect all over the plant by mimicking the natural hormone. Narrow-leaved monocotyledonous plants are not affected. By contrast, recently developed chemicals will act against a grass weed growing in a grass crop. Chemicals such as Barban can prevent the growth of wild oats in, for example, wheat. This chemical interferes with cell division in the susceptible plant.

Selective herbicides have increased in complexity and number over the years and we may expect this progress to continue; specificity and protection for the crop can never be too precise.

How does a herbicide act?

A *contact herbicide* acts only on the regions of the leaf which receive it. If the spray liquid does not wet the whole leaf, scorch spots develop.

Figure 5.10 The effect of a selective hormonal weedkiller (J.B. Land).

Sulphuric acid, for example, destroys the cell membranes and cell contents. Paraquat is far more specific; it interferes with the mechanism of photosynthesis (by accepting the energy trapped by chlorophyll) and destroys all membranes.

The *selective herbicide* must be taken in by the weed and transported to the site of its action. Foliar sprays are applied under 'good growing conditions' because the chemical will then be transported rapidly round the plant. During slow passage in dull, cold weather, the herbicide may be destroyed by the plant. 2,4-D travels the same route as sugar; it moves along the cytoplasmic connections between the leaf cells (the symplasm) and then enters the phloem transport system. The residual herbicide Simazine, on the other hand, travels from the soil through the non-living cell walls to the xylem (the apoplast system), and thence to the leaf via the transpirational stream. This movement is quite passive; the plant cannot help taking up the herbicide.

Once in the plant, 2,4-D interferes generally with control of growth (Figure 5.10) and specifically with cellular respiration and the synthesis of fats and proteins. The action of Simazine is more specific; it inhibits that part of photosynthesis in which light energy is used to split water molecules (and interferes with protein and nucleic acid synthesis).

A new herbicide, **glyphosate**, which shows promise for the 1980s is based on the amino acid glycine. Its reported features make it an ideal total herbicide: it is readily translocated and so has systemic action, it is broken down to harmless products by micro-organisms in the soil in a few days, it has low mammalian toxicity and it acts on an enzyme system unique to plants. Because it is translocated, small, local applications of herbicide solution to leaf or cut stem ultimately kill the plants. With the knowledge of the biology of the herbicide, nature reserve wardens have explored the use of this chemical to eradicate single species; applications via a rope wick at the end of a tube-like reservoir for the solution have been successful in controlling weeds such as ragwort and thistles in some nature reserves. This use contrasts markedly with the way the farmer uses the chemical to create a strip of soil devoid of both perennial and annual weeds around his much valued cereal crops.

As we learn more about the *normal* structure and function of plants, we can identify more precisely the effects of herbicides at the cellular level. We can also establish their action in other organisms in food chains, of which these plants form part, and their persistence in soil and in organic matter – knowledge which is essential if we are to understand both the specificity of these chemicals and their effects on other members of the ecosystem as the basis of their intelligent use.

Agents of disease competing for food

Between host and parasite there is competition for building blocks and energy, competition which causes an enormous loss of food production. The effect of a disease depends upon the genetic background of the host, which, as we shall see in Chapter 7, we can modify. Because of this, and because disease is so important, discussion of it forms a separate chapter (Chapter 10).

Diets for animals

Obtaining the highest possible yield per individual is usually of more importance when rearing animals than when growing plants. Traditionally it has been difficult to measure the cost of animals' food and farmers have known their animals as individuals. Every effort is made to maximise the conversion of low-grade food into high-grade food, but the farmer must remain aware of the implications of the law of limiting factors. Two examples will serve to illustrate this.

The first example is that of a single simple factor missing from the diet – a trace-element deficiency may have disastrous effects. Deficiencies of copper result in a disease called sway back; low iodine levels give goitre and low cobalt levels cause pine in sheep. Cobalt deficiency is of particular interest because ruminants are more susceptible than animals with post-gastric fermentation (such as horses). This difference is not understood, but we do know that cobalt is an essential component of vitamin B_{12}, which in turn is required for the proper formation of red blood cells; deficiency of vitamin B_{12} in humans causes pernicious anaemia. A mere 2 kg of cobalt sulphate added per hectare of grassland every three years will prevent any deficiency symptoms in cows and sheep, yet when this small quantity is missing from the diet, the effect is so drastic that the animals may waste away and die.

The second example concerns the nature of the whole diet rather than an isolated factor. This may be ill-suited to the animal's digestion, absorption and utilisation processes. For example, it is wasteful to feed large quantities of plant fibre to non-ruminants. Knowing the characteristic ways that particular animal species use different foods, the composition of their diet can be adjusted so that the best use is made of both food and animal. As with the iron deficiency discussed on page 55, the animal–diet relationship must be clarified. When the interactions are understood, biological solutions to feeding problems can be predicted.

The futility of feeding cellulose to a pig was clearly shown in the early 1940s by experiments carried out in the UK during food shortages. Pigs were fed an experimental diet containing 85 per cent fibre with protein supplements. Animals weighing about 70 kg used almost 3 kg of this food per day and grew by 0.5 kg per day. The control group of pigs, reared on a 90 per cent cereal diet (providing readily digestible starch) and protein supplements, ate only 2 kg of food but grew at the rate of 0.7 kg per day.

This example is extreme, yet there is a real problem in formulating a balanced diet for pigs and poultry. Unlike ruminants, but like humans, they have precise amino acid requirements (see Chapter 2). Although the total quantity of protein in pig or poultry food may be adequate, the amount that the animal can *use* depends on the proportion of essential amino acids in the food. Like humans, the animals do not thrive when the proportions are wrong, but grow well when they are correct – the 'balanced diet'.

Maximum growth rate, then, does not necessarily demand the use of expensive foods but rather the intelligent use of available foodstuffs. The farmer must be sensitive to the costs of foods and to their use by particular animals.

The number of organisms on an area of land

Plants are stationary: when we grow similar crop plants in an area of land, they compete with each other for light and nutrients as they compete with weeds. The competition may be intense, because they exploit the same regions of the soil and of the air above it. Isolated plants usually grow and crop best; at a certain density, the yield per plant begins to fall (Figure 5.11). The yield of total dry matter per unit of land rarely declines at high densities, but the *edible* part of that yield may fall. So we try to grow plants spaced so that the field gives the maximum yield for the least number of seeds.

If the seed is very expensive one year, farmers may be advised to reduce the planting density, accept a slightly lower yield, and spend much less on seed. The cost of 'seed' potatoes (really selected stem tubers) fluctuates from season to season; as farmers need to plant about 80 kg/ha, they tend to adopt the policy of planting for profit rather than yield.

Animals were traditionally reared under conditions which maximised their output, but this is changing; under certain circumstances,

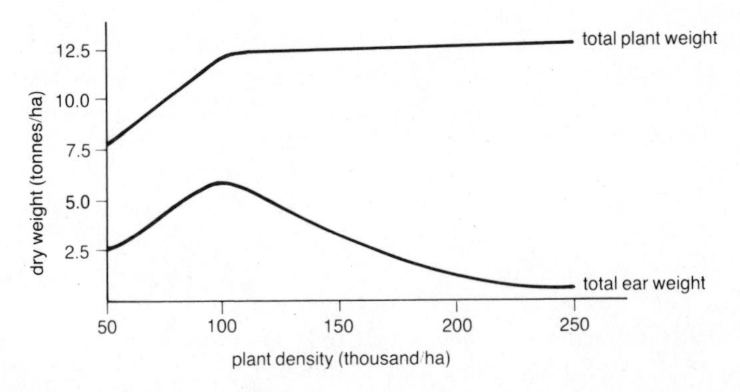

Figure 5.11 The relationship between plant density and yield of maize. (After Eddowes.)

the farmer may be more concerned with maximum output per unit of land, per unit of farm buildings, per unit of capital invested, per unit of work force, or per unit of food. This is particularly true of intensive systems, as we shall see in Chapter 9.

So we return to the law of diminishing returns; the cost of improvements to the feeding environment must always be set against the value of the improved yield.

Summary

Ideally, plants and animals should be reared with an abundant supply of all the components of their feeding environment required for their growth. The producer must be aware of the practical and economic aspects of ensuring this. Two options are available:

a *additions* may be made to the environment to mitigate the effects of the law of limiting factors;

b organisms which *compete* for nutrients may be removed from the environment.

6 Producing excess biomass – growth and reproduction

When we eat, we remove organic material from the food chain; our continual need for food generates a continual need to replace this biomass. Chapter 5 showed how this can be facilitated by growing organisms under favourable conditions; Chapters 7 and 8 discuss ways of altering the food producers genetically so that they will make food – but not necessarily biomass – more efficiently. This chapter considers the production of biomass, the growth pattern of individuals and their use as food when their feeding has been most efficient.

In order to increase and to control the rate of replacement of organisms which we eat, we must understand the way in which individuals reproduce to give new individuals. Rates of reproduction vary widely – the cow has a generation length of two and a half years, for example; a micro-organism perhaps twenty minutes. Reproduction rates restrict the production of food from animals, but rarely from plants. When micro-organisms are grown to produce food (Chapter 11), and when they depress the yield of other food-producing organisms (Chapter 10), they rarely reproduce at the maximum possible rate; their feeding environments affect this potential. In most farm animals we can control the time of reproduction and to some extent the number of offspring. In plants, the great rate of *sexual* reproduction is enhanced by methods of asexual reproduction.

Food-producing cycles

To keep food available to us, there must be an **excess of biomass** in the food chain; a fine balance exists between the production of organic material and its use. An organism's capacity to reproduce is one factor which restricts the rate of increase of biomass; reproduction implies an increase in numbers.

It can take place by two distinct patterns of **cell division**. The first, **mitosis**, allows a cell to replicate itself, to produce two genetically similar cells after a period of synthesis. If these cells stay together, the biomass increases when cells multiply within an individual. This type of cellular reproduction is shown, for example, by a fertilised cow egg, which progressively divides (and differentiates) into a calf and finally a mature animal. If the two cells become separated and form independent organisms, there is an increase not only in biomass but also in the number of individuals. This is the pattern of growth of a single-celled organism such as a bacterium or yeast.

Some plants – complex multicellular organisms – reproduce by mitosis alone, as is shown by grafting, taking runners or using tubers to propagate **asexually** or **vegetatively**; in each case, a mass of differentiated cells gives rise to a new individual. The essential property of mitosis is that genes are replicated exactly without the planned opportunity for change, so the genetic constitutions of all individuals in a population of yeast cells are the same, as are those of all the cells in the mature cow and of all the individuals obtained by vegetative means from an apple tree.

By contrast, higher animals and some plants reproduce only by sexual methods, which use the second type of cell division, **meiosis**, to produce the gametes which characterise **sexual reproduction**. Meiotic cell division allows reassembly of the genes which will give rise to the new organism; later chapters will discuss how this is exploited in breeding programmes.

This type of reproduction takes place only in organisms approaching physical maturity, so reproduction occurs when the 'body' of the organism is ready for use as food. There is thus a conflict between the need for food and the need for replacement.

Patterns of growth

It is difficult to define 'growth', yet growth is the basis of food production. Put simply, **growth** is an irreversible increase in size; simple parameters of growth include measurements of height or length or cell number. But these do not account for the change in complexity seen when a seed germinates, when a larva pupates, or when we ourselves grow. When the proportion of structures within the organism changes, growth is said to be **allometric**. Increase in size may be the result of water uptake – shown by imbibition in seeds and by fruit development – so **dry weight increase** is often chosen as the best measure of growth.

Different organisms grown under ideal conditions increase dry

Table 6.1 Rates of dry weight increase in a range of organisms.

Species	Maximum % dry weight increase over 24 hours
Escherichia coli	4750
Yeast	1400
Sunflower	29
Scots Pine	5
Cow (as foetus)	1

weight at widely varying rates. Table 6.1 gives some examples. Complex organisms generally grow much more slowly than simple ones; this is because life is sustained by a high level of structural complexity, and systems such as transport make demands on the organism at the expense of its growth rate.

Micro-organisms have enormous growth potential, which is important in two ways. Firstly, many disease agents fall into this group, and microbial growth rates may therefore affect the spread of an epidemic (Chapter 10). Secondly, micro-organisms or their products may be used as food, so their growth rates affect the efficiency of food production.

Growth of unicellular organisms

Unicellular (single-celled) organisms grow explosively: each and every cell may divide to form two daughter cells as little as twenty minutes after the cell itself has been formed. Before modern antibiotics were available, harmful bacteria grew in our bodies so fast that they soon overpowered our defences.

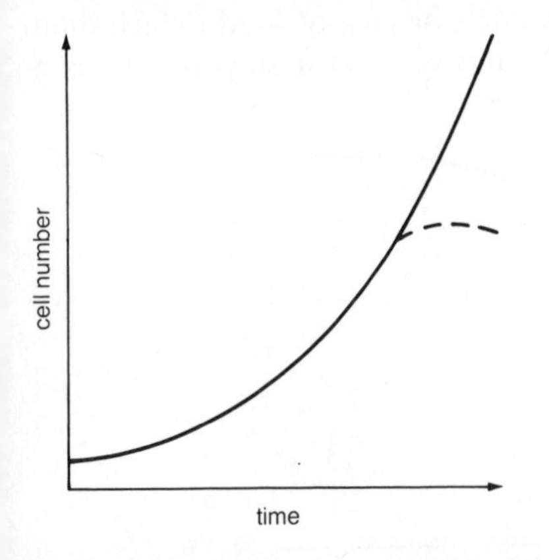

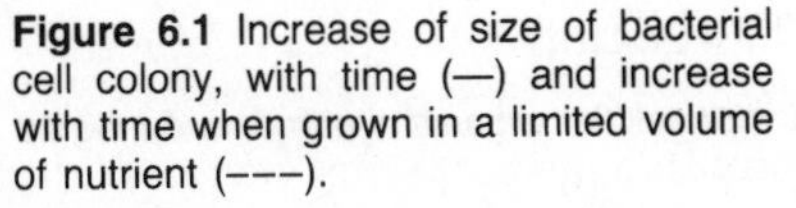
Figure 6.1 Increase of size of bacterial cell colony, with time (—) and increase with time when grown in a limited volume of nutrient (---).

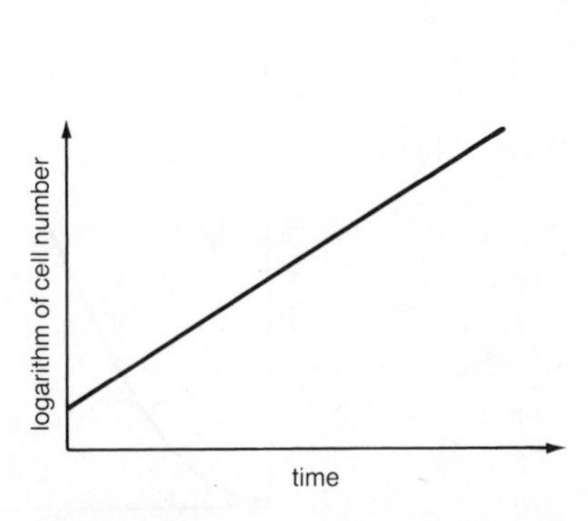

Figure 6.2 Increase of size of bacterial cell colony (Figure 6.1) with time: use of the logarithm of the cell number gives a straight line.

Growth in these single-celled organisms is often measured by counting their number in a certain volume of culture medium. There are cells at all stages of development, so a sample will record the number of 'average cells'; and this number is proportional to their average dry weight.

In a simple multicellular organism, or in a given population of unicellular organisms, the number of cells during growth increases at an accelerating rate, because the number of daughter cells depends upon the number of cells already there (Figure 6.1). If we plot the logarithm of the number of cells against time, we get a linear relationship (Figure 6.2): hence the term **logarithmic** or **exponential growth**.

When energy and nourishment are less available, and when toxic waste products accumulate, the rate of increase falls, and gradually the number of cells becomes constant; the culture is then said to be **stationary**. This sequence of rates gives the characteristic **S-shaped** or **sigmoid** type of curve (Figure 6.1). Eventually, the number of cells may even decline; we *encourage* this in disease control (Chapter 10), but we try to *avoid* it in fermentation (Chapter 11).

Growth of multicellular organisms

Growth of a multicellular organism can be measured by estimating its dry weight, but this involves killing the organism, and so non-destructive measures of growth – the length of part of the organism such as a bone, the number of leaves or the fresh weight – are often used instead. The growth pattern obtained thus (Figure 6.3) is also sigmoid. Here, though, it is the genetic make-up of organisms which determines their final size, rather than toxic products or lack of food (which limit microbial growth). Little is known about why a mouse stays the size of a

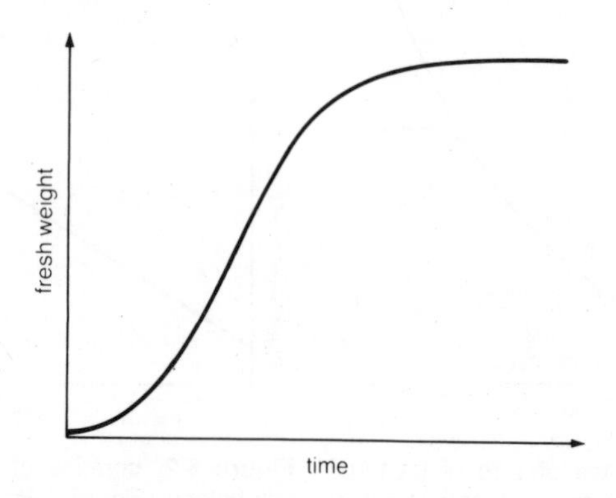

Figure 6.3 Increase of the fresh weight of a multicellular organism with time during its growth.

mouse, whereas a cow grows relatively large, or why wheat plants rarely grow taller than 100 cm while maize may reach 2.5 m: for now we simply accept and exploit this.

Two aspects of the S-shaped growth curve are important in food production. Firstly, young organisms grow slowly because the feeding organism is small. Secondly, the old become mature and contribute to reproduction rather than biomass production. The variable speed of growth can be described by plotting against time the *increase* in the growth parameter over a given period of time, e.g. kg addition per day, as the absolute growth rate, rather than the cumulative value which would be recorded as, say, the actual body mass on that day. But since our concern is with surplus biomass available as food, we should consider growth in terms of the biomass needed to produce it. Thus we derive the **relative growth rate (RGR)** (see Figure 6.4), obtained by

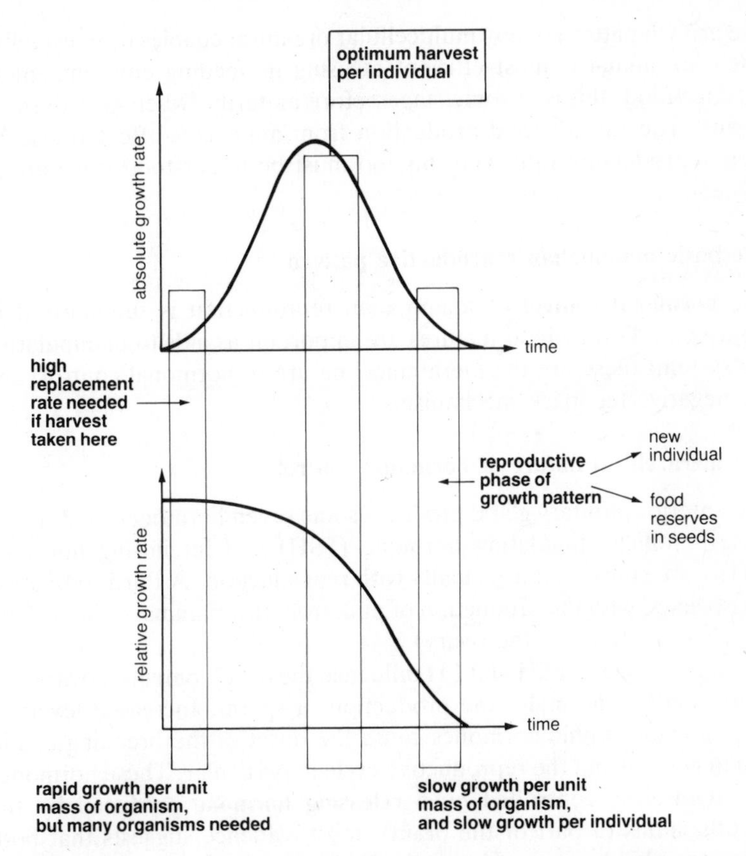

Figure 6.4 Absolute and relative growth rates of individuals, showing the effectiveness in food production of the various stages of growth.

plotting, say, addition to body mass as a proportion of the body mass which produced it over a given period of time; this shows us that growth rate in young organisms far outstrips that in older organisms of the same species. So it is that we aim to use young individuals for food; but they must not be so young that all our efforts are directed towards producing new individuals. As a note of caution (taken up again in Chapter 9), this argument assumes that we are interested in the *whole* organism rather than just a part of it: this effect of partitioning means that the young organism may not be useful as food (for example, young cereals do not have ears of grain) and it is often necessary to wait for the mature organism.

Reproductive patterns in mammals

The growth pattern of any multicellular organism enables us to establish when an animal is most efficient at using its feeding environment to produce food; this is at some stage before maturity (when growth slows down). The rate of food production from animals is affected also by their reproduction rate, and this too must be understood if we are to influence it.

The basic mammalian reproductive pattern

The **hormonal control** of mammalian reproduction is summarised in Figure 6.5. Two general features are important as aids to manipulating the system; these are the hierarchical nature of hormonal control, and the negative feedback mechanisms.

The hierarchical nature of hormonal control

The anterior pituitary gland produces some seven hormones; and two of these – follicle-stimulating hormone (FSH) and luteinising hormone (LH) – are concerned principally with reproduction. A third, prolactin, is associated with the production of milk from the mammary glands (and has also an effect on the ovary).

In the female, FSH and LH influence the development of ovary and ovum; and in the male, the production of sperm. Increased levels of these gonadotrophic hormones cause the onset of the breeding condition in general and the reproductive cycle in particular. These hormones are themselves controlled by a **releasing hormone** produced by the hypothalamus (a part of the brain); most evidence suggests that both gonadotrophic hormones are controlled by one hormone, called gonadotrophin-releasing hormone (Gn-RH). Gn-RH has been purified and

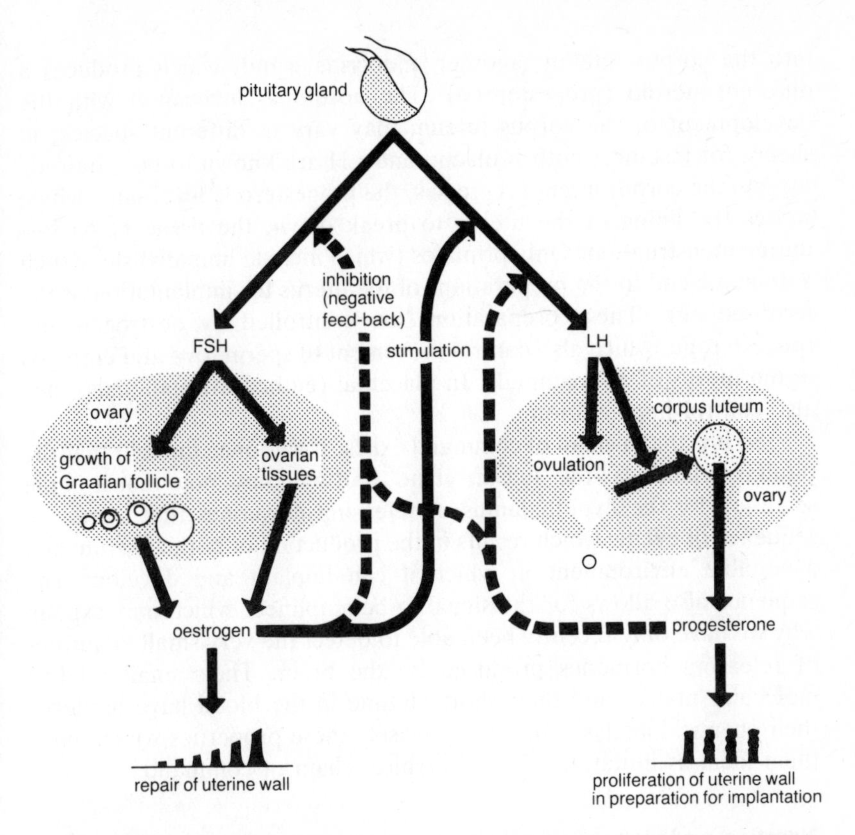

Figure 6.5 Hormonal control of the female reproductive cycle. Solid arrows signify stimulation, broken arrows inhibition.
(After Roberts, 1982, *Biology, a Functional Approach*, 3rd Ed., Thomas Nelson.)

proves to be a relatively simple chemical, made of just ten amino acids (a decapeptide). It is much smaller than FSH or LH, which comprise polysaccharide–protein complexes; these glycoproteins have a **relative molecular mass** (**M_r**) of about 50 000. Gn-RH is produced in the cell body of a nerve cell in the hypothalamus, and passes to the pituitary gland by a local portal system. Gn-RH release may be stimulated by a fall in the level of ovarian hormones which had hitherto inhibited its production, or by some external stimulus such as day length which changes with the season. Both FSH and LH influence the ovary. High FSH levels help the follicle to develop and, with LH, stimulate the ovary to produce a steroid hormone (oestrogen). Thus the follicle is both the source of the gamete and an endocrine organ in its own right. A surge of LH causes ovulation, and the remains of the follicle in the ovary develop into the corpus luteum, another endocrine gland, which produces a different steroid (progesterone). The hormones associated with the development of the corpus luteum may vary in different species; in sheep, for instance, both prolactin and LH are known to be required.

As the corpus luteum regresses, the progesterone level falls, which causes the lining of the uterus to break down, the tissue being lost during menstruation. Only primates (which include humans) show such a dramatic end to the preparations of the uterus for implantation of the fertilised egg. These preparations are controlled by oestrogen and progesterone, which also control movement of sperm, ova and embryos in the uterus of all mammals. In placental (eutherian) mammals, they also control implantation.

So it is that a chain of command exists. It was once thought that the pituitary gland was the 'master gland', but we know now that it is the secretion by the hypothalamus of releasing hormones that starts the sequence of events which results in the production of both an ovum and a suitable environment in which it can implant and develop. The sequence also allows for the signal to be amplified, which may explain why we have only recently been able to detect the very small quantities of releasing hormones produced by the brain. Their small relative molecular masses and their short lifetime in the blood have hindered their chemical analysis, but it is precisely these properties which make them ideal as initiators of a heirarchical chain of command.

Negative feedback

When the product of a process inhibits the reaction which makes it, the rate is said to be subject to **end-product inhibition** or **negative feedback inhibition**. The mechanism is illustrated by the interaction between FSH and LH with the ovarian hormones.

Gonadotrophins (FSH and LH) produced by the pituitary cause the ovary to secrete oestrogens, other steroids and inhibin. We find that oestrogen – needed at fairly high levels to initiate the surge of LH and ovulation – controls its own production; and oestrogen with other gonadal hormones inhibits FSH and LH production. So fluctuations in the production of gonadal hormones are regulated by two methods. On the one hand, lower levels of FSH result in lower levels of production of gonadal hormones; on the other, higher gonadal activity reduces, by negative feedback inhibition, the level of FSH and LH in the blood.

Negative feedback is a basic control mechanism applicable equally to the most minute cellular activity and to the activity of an organ, organism or population.

Intervention in mammalian reproduction

With the same aim of optimising food production, two options are available: the reproductive cycle can be manipulated to hasten the production of new individuals, or different animals can be selected on the basis of their different breeding characteristics. Manipulations with existing animals may prove costly; and reproductively superior animals may not produce the right quantity or quality of food.

Environmental control of time of ovulation

Because short days remove the inhibition of FSH and LH production in some seasonal breeding animals (such as temperate sheep and goats), these tend to breed at one time of the year. They can be induced to ovulate early by bringing the animals into buildings and decreasing day-length artificially. The critical feature of this treatment is the change in day-length rather than the absolute number of hours of daylight, so the return of the animals to normal day-length after a period of artificial lighting to *extend* the day-length can provide the necessary stimulus for ovulation. This requires the animals to be kept in a building under careful control, and hence is most suitable for intensively reared animals.

Hormonal control of the time of reproduction

By introducing progesterone into the animal, we can control the **oestrous cycle** and so induce a breeding condition (**oestrus**) and ovulation at a predetermined time. Progesterone prevents ovulation by blocking the positive feedback of oestrogen on the pre-ovulatory discharge of LH. Progesterone is secreted from the corpus luteum in an animal which has just ovulated, and from the corpus luteum and the placenta of a pregnant animal. Because it prevents the surge of LH from taking place, the Graafian follicle in the ovary is prevented from developing beyond the pre-ovulation stage.

The hormone itself can be given by injection into the animal's muscle, injections being repeated over several days to make sure that the natural corpus luteum from the previous ovulation has regressed fully. After the last injection has been given, the blood levels of progesterone fall; sheep and cattle return to oestrus in about two days. Ovulation in sheep can be synchronised; the corpus luteum is normally maintained for about twelve days, so treatment with progesterone of a group of animals for at least twelve days ensures that their corpora lutea no longer produce any progesterone. Progesterone is a fairly short-lived hormone (its half-life is about half an hour) which accounts for the speedy and precise return to reproduction. However, because of work carried out to develop chemicals for oral contraceptives in women, we now have progesterone-like hormones, **progestagens**, which have half-lives much longer than progesterone. There is now a subtle method of administering these hormones. The progestagens are soaked into polyurethane foam blocks and the blocks placed in the vagina, where the hormone is slowly absorbed through the mucosal lining; when the blocks are removed, the effect is the same as that of stopping the course of injections with progesterone. The hormone is absorbed because being a steroid, it is fat-soluble and will enter cells through the lipids of the cell membranes. (Steroid hormones administered by mouth are effective, whereas protein hormones such as FSH and LH – and insulin for that matter – are broken down by the animal's own digestive enzymes.)

Use by humans of progestagen-containing oral contraceptives is now accepted, but what about their use with other animals? Most owners of bitches consider as rather a nuisance the period when their animal is on heat. How much nicer if the socially inconvenient display of oestrus could be suppressed! An obvious method is hormone treatment. From the commercial viewpoint, the ideal might be to add a progestational agent into a particular dog food, thereby preventing oestrus; should the dog owner revert to buying a non-progestagen treated food there would be a resumption of the physiological events which lead to oestrus. Once he started feeding the treated food, the dog owner would be 'hooked' on the product! Such an option is not in fact available either to the food manufacturer or to the dog owner, because of legislation regarding food standards – *all* pet food must be fit for human consumption.

There is another 'time-control' option. Prostaglandins, which are made by the body from an 'essential' fatty acid – arachidonic acid (page 14) – are concerned with the natural regression of the corpus luteum. Prostaglandin F-2α is particularly important and may be used to induce the regression of the corpus luteum directly in cattle and sheep. The half-life of these hormones is particularly short (a matter of a few minutes); they are removed from the blood as it passes through the lungs. Pharmaceutical research for similar compounds has shown that they also induce the contractions of the uterus which precede birth or abortion. One injection of these artificial prostaglandins causes the corpora lutea to regress and the animal returns to heat two days later.

At a particular stage of the oestrous cycle (around the time of ovulation), there is no corpus luteum to regress; but synchrony is achieved in a group by means of two injections, separated by the correct interval (twelve days for cows). By the time of the second injection, each individual has a corpus luteum so all ovulate together.

Hormonal control of the number of eggs available for fertilisation

That the number of eggs shed by the ovary at ovulation might be increased is an attractive possibility. So attractive is it that this might have been our major method of manipulation; more eggs should mean more embryos, more young and hence potentially more units of food production. One of the many snags to this idea is parallelled by the problems which faced gynaecologists helping women to conceive. Using gonadotrophins, it is difficult to control the number of eggs produced at ovulation; this generally results in multiple births. A frequent result is that the babies are small and premature, and few survive. So it is with farm animals. Nevertheless, if gonadotrophic hormones are injected into animals then there is an increase in the number of developing follicles. Hence at ovulation, many eggs are shed. The same problems of controlling the numbers and mortality of the newborn exist as with humans.

Another problem is that the higher levels of oestrogen produced by the ovary (because of the high gonadotrophin levels) can seriously affect the movement of sperm and ova in the Fallopian tube, even to the extent that fertilisation does not take place.

Because of these problems – high infant mortality and reduced chance of fertilisation – scientists are seeking other ways of encouraging animals to produce more young. The likely avenues of investigation will include the use of drugs to inhibit oestrogen action, or immunisation against oestrogen; because it is this hormone which inhibits the production of FSH, and FSH which controls the number of follicles which develop in the ovary.

Indeed, the medical approach to infertility may give us clues to the future of controlled ovulation in farm animals. We hear much less about multiple births in women these days; this is not because we no longer attempt to help infertile couples nor because we have grown accustomed to these events. There is a real decrease in multiple births, due in part to two recent advances. Firstly, we can actually measure the oestrogen level (which correlates with activity in the ovary) in the female's blood; this information enables us to estimate the number of follicles developing. The quantity of supplementary hormones can then be calculated precisely. Secondly, drugs such as clomiphene (which is a weak oestrogen) may be used to reduce the feedback effect of oestrogen. Evidence suggests that the drug binds to oestrogen receptors in the hypothalamus but does not stimulate these receptors (Figure 6.6).

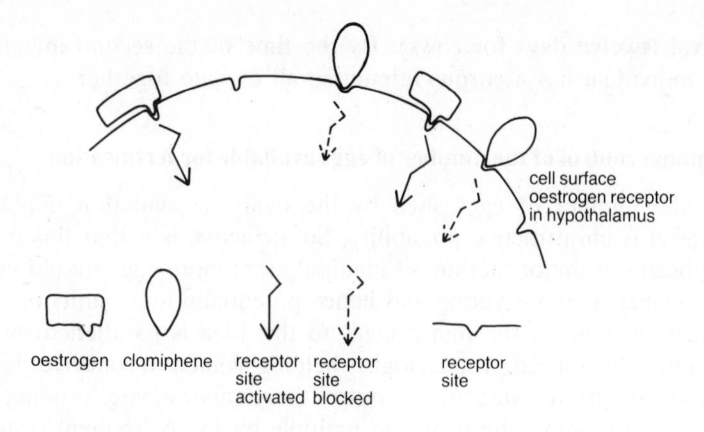

Figure 6.6 One way in which the oestrogen inhibitor might act.

Fewer receptors are therefore available to the oestrogen produced by the follicle. The hypothalamus interprets this as a low oestrogen level and secretes more gonadotrophic hormones. This exploits the natural response of the system. Use in humans is analogous to adjusting the faulty indicator of a heater's thermostat to read the correct value; use in farm animals is like setting the thermostat to a higher level.

Reproductive potentials of different breeds – genetic differences

Breeds differ in their reproductive capacity, and scientists are investigating how these natural differences may be exploited during breeding programmes. Consider the example in Table 6.2.

Table 6.2 The reproductive capacity of sheep.

Breed of sheep	Age at first lambing (years)	Average litter size
Finnish Landrace	1	2.7
Tasmanian Merino	2	1.0

Compared with the Tasmanian Merino, the Finnish Landrace has considerable reproductive potential because it produces more young when it does reproduce and can reproduce at an earlier age. Yet we do not rear Finnish Landrace as a pure breed because of other disadvantages, while the Merino is kept as a highly productive wool-producing breed.

Nevertheless, highly reproductive breeds can be a valuable source of material for breeding programmes, and scientists try to exploit this reproductive merit in such programmes.

Other effects of reproductive hormones – castration

Before leaving mammalian reproduction, we must mention the contribution to food production made by castrated male animals. Farmers castrate males of most domestic animals for three reasons, all of which stem from the removal thereby of the source of male hormones (including testosterone) so that their aggression is suppressed. Firstly, they do not dissipate energy in seeking out a mate and fighting other males, so more of their food is used to make them grow. Secondly, they are much more docile to handle in confined spaces and so lend themselves to intensive methods of production. This is at the expense of their growth-rate characteristics; they grow less rapidly than their intact counterparts, and their carcasses have more fat (although castration late in life can make for better carcass quality). Thirdly, the housing of intact male cattle is governed by legislation; in Britain, demands for secure fencing are so severe that the overall economics dictate castration. This is not the case on mainland Europe where the balance tends to be the other way; legislation is less stringent and low-fat meat is in great demand. About 90 per cent of male cattle in the Federal Republic of Germany are reared intact.

Reproductive patterns in plants

There is a contrast between the variation shown when plants reproduce sexually and the constancy shown when reproduction is asexual or vegetative. Variation is exploited in breeding programmes (Chapter 8) because the products of meiosis confer differences upon the progeny. Vegetative propagation, by contrast, exploits the accurately replicated products of mitosis, and results in uniformity favoured by the agriculturalist (Chapter 4). Vegetative propagation can also be used to speed up reproduction in non-annual plants, hence many crops can be brought into cropping much more quickly.

The sexual cycle in higher plants

The haploid nuclei which result from meiosis are contained in pollen grains (from the anther) and in the embryo sac (within the ovule). Unlike the animal system, each nucleus follows this division by mitosis (Figure 6.7), forming the much reduced haploid male and female gametophytes. When a nucleus from the pollen fuses with one from the embryo sac, the diploid number is restored and fertilisation has occurred. Thereafter mitotic cell divisions allow the embryo to develop.

Pollination is simply a transfer of pollen from the anther to the

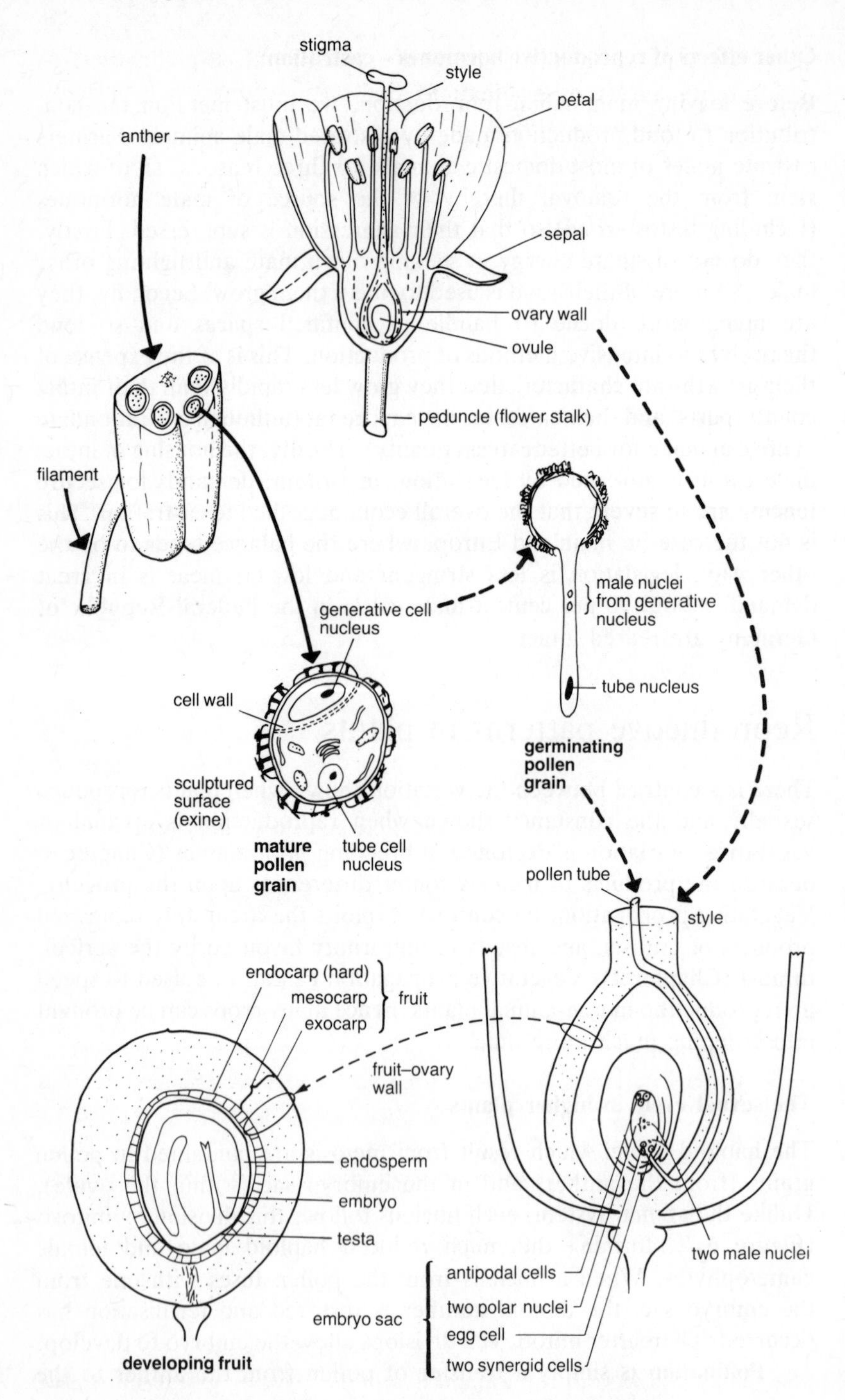

Figure 6.7 Sexual reproduction in a flowering plant such as the cherry.

stigma; anther and stigma may be on the same plant (**self-pollination**), or on different plants (**cross-pollination**). Self-fertilisation is likely to follow self-pollination, cross-fertilisation to follow cross-pollination.

Intervention in plant reproduction

Products of fertilisation – our food

Rarely do we eat the whole of a plant. When all the parts above the soil are used, we usually discard the roots. Even when plants are grown for animal food there is some waste, despite the fact that they often eat much more of the available biomass. This phenomenon of **partition**, where only part of the organism is used for food, will be discussed in Chapter 9; we now trace the development of some reproductive structures which we use as food. These structures are: the fruit, which is the **ovary wall**; the **embryo**, which is produced by fertilisation and may ultimately develop into a new plant; the **endosperm**, formed when a generative nucleus from the pollen tube fuses with two nuclei in the embryo sac; and the **receptacle**, which carries the parts of the flower, and the **flower head**.

The ovary wall and receptacle can be considered together, both comprising entirely maternal tissue so that they reflect the genetic make-up *only* of the egg-producing parent. Both may become fleshy and succulent, which visual similarity leads to confusion over their origin and development. We eat this tissue of apples and pears (receptacle) and almost any fleshy vegetable or fruit – aubergine, pepper, cucumber, marrow, squash, tomatoes and all the berries. Flower heads (or inflorescences) give us such foods as cauliflower and fig.

The embryo provides the species with its continuity from generation to generation; if it develops extensively, we may use it as food. The endosperm, the other product of cell fusion, is used for storage; the embryo may use this food when the seed germinates, as it does the stores in the cotyledon – unless we harvest it as food. Seeds generally may be described as either endospermic (such as cereals and castor bean) or cotyledonary (such as legumes).

The vegetative structures – stem in celery, potato, Jerusalem artichoke, sugar cane; leaf in spinach, cabbage, lettuce; root in carrot, parsnip, turnip – all develop following the growth of the embryo into a new individual.

Control of reproduction

For plants to reproduce, the bud and its growing point must stop producing leaves and instead produce a flower; the vegetative apex must change to the floral apex. The flower parts are modified leaves, so a change in the developmental processes is required. These are not fully understood, but it is known that there are two factors which initiate them. Firstly, it is mature plants which produce flowers. Plants pass naturally from a vegetative phase into a floral phase as they age; thus many annual crops (and the weeds which grow among them) produce flowers and later seed towards the end of their growing season.

Secondly, many plants start these reproductive changes after a signal from the environment. Such signals are important in perennial or biennial plants; two known signals are cold treatment (**vernalisation**) and changes in day length (**photoperiodism**).

The photoperiodic effect was first noted by Allard and Garner in tobacco plants, when they tried to grow the particular northern strain of tobacco (*Nicotiana tobacum*) 'Maryland Mammoth' in the southern states of the USA. They hoped that growth and flowering would be faster in the warmer summer of the south, but they observed the opposite; plants grew well, but failed to flower; the strain needed the shorter days of the north to initiate the flowering process. That we now know much more about photoperiodism illustrates the way in which scientific research is often fostered by accidents.

Sugar beet flowers as a result of cold and long days. The plant is a biennial; after germination it grows vegetatively in its first year (producing a large sugar-rich tap root and a rosette of leaves), then flowers in the following year. The root is harvested at the end of the first year when all the sugar reserves can be utilised to make table sugar. The yield of roots depends largely on the time for which the plant has been growing. Naturally the farmer hopes for high yield, so sows seeds early in the year. But because the seed of cultivated plants have no dormant period (Chapter 4), growth gets under way early; the young plant is then exposed to environmental factors which would, for a wild plant, initiate flowering. The cold periods of the early part of the growing season contribute to the vernalisation effect while the longer days of summer contribute to the photoperiodic effect. The farmer gambles; he sows early for a high yield, accepting that environmental stimuli may 'persuade' the sugar beet plants to behave as annuals and dissipate their sugar reserves by flowering (bolting).

By contrast, winter varieties of wheat (*Triticum*), rye (*Secale*) and barley (*Hordeum*) remain vegetative unless exposed to the chilling effects of a temperate region winter. So if grown in a tropical region, or sown in spring rather than winter, there is no possibility of them yielding grain. Spring varieties, which give lower yields than their winter counterparts, do not need cold treatment to initiate this partition of the biomass.

Seed and fruit development

The development of the seed usually depends upon fertilisation. The fertilised egg, and later the embryo, produces auxin, a chemical message which stimulates the ovary wall or receptacle to keep this tissue alive and allows it to develop. Pollen grains themselves contain auxin (indole acetic acid or IAA). Figure 6.8 depicts (for strawberry and apple) what happens when some of the ovules are not fertilised. Not surprisingly, fruit growers are anxious to encourage bees and other pollinating insects in their orchards during the flowering period!

Scientists have been able to mimic the effects of fertilisation. A synthetic chemical, naphthalene acetic acid (NAA), is substituted for the natural one and ensures ovary or receptacle development. This process of **parthenocarpy** is used by tomato and pear growers to overcome poor fertilisation following bad pollination (Figure 6.8). It is sprayed on to the flowers so that the fruit develops – with one difference, however: the fruit is pipless!

Asexual reproduction

Vegetative methods of propagation which utilise plants' natural method of reproduction have fairly modest rates of increasing numbers. Runners (strawberry), stem tubers (potato) and root tubers (sweet potato and cassava) are all used by a plant breeder to 'bulk up' a new strain once it has been decided to make it commercially available, but he may expect this to take about three years.

New individual plants can also be produced from smallish pieces of old ones. Hence the process of taking cuttings, whereby a small section of the stem or, in some cases, the root, is 'planted' under favourable conditions. If the cutting contains enough tissue to serve as a food supply and is in a moist (but not wet), warm soil, the mature tissue forms a mass of undifferentiated cells (callus) at the cut surface. This callus then differentiates to form roots which support the original cutting. NAA can again be used to encourage this process with plants such as sugar cane, blackcurrants, pineapple, cassava and many ornamental plants. But the hormone treatment only *encourages* roots to develop, by supplementing the natural levels of auxin and so increasing them in proportion to the level of kinetin, a plant hormone more concerned with cell division; the change in hormone balance speeds up rooting. As we understand the plant–environment interrelations more fully, treatment may prove unnecessary.

Grafting also provides us with a rapid method of propagation. Here a small portion, generally a bud or a short shoot (the scion), is attached to a rootstock so that the cambia of each touch; continued division pro-

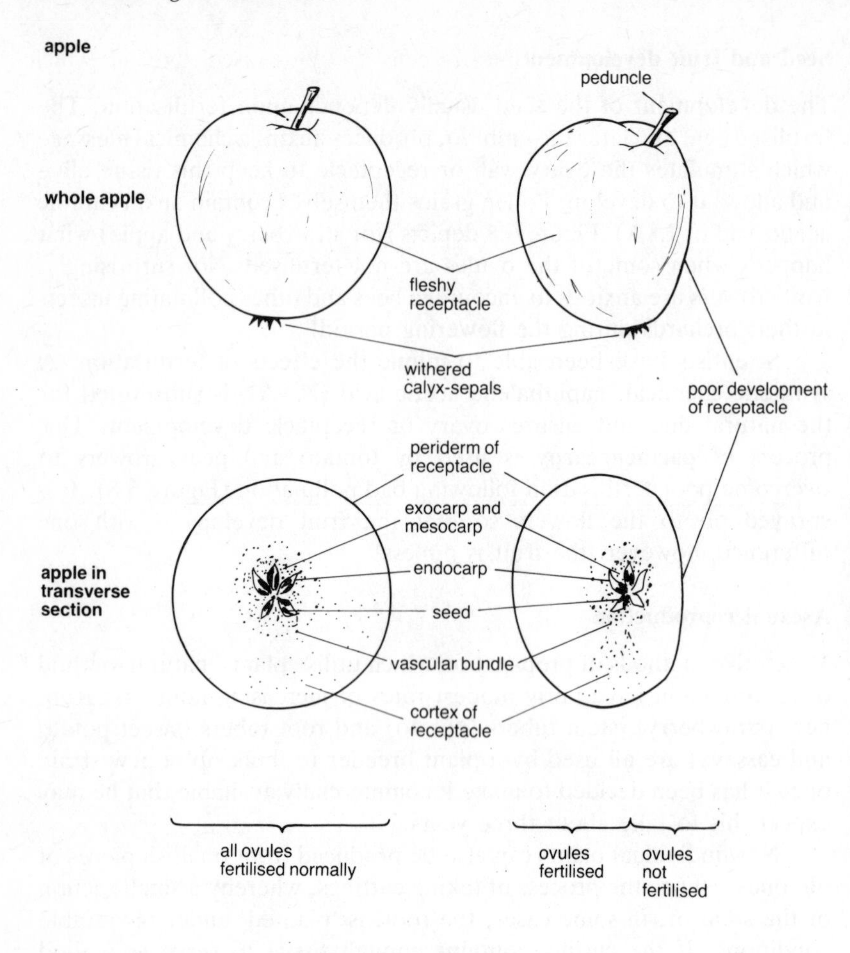

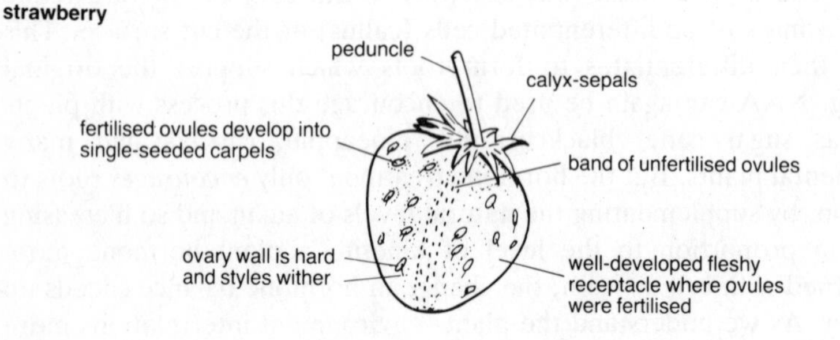

Figure 6.8 The effect of ovule infertility on the development of the fleshy false fruit of strawberry and apple.

duces a callus, from which mass of cells develops vascular tissue, which joins the two plants (Figure 6.9). There are several methods of making a graft, but two will serve to illustrate the principle (Figure 6.10).

The rootstock may be just a wild ancestor of the food-producing plant, but we often find that as much care and research has been devoted to breeding the rootstock as has gone into testing the suitability of the scion. A range of rootstocks may exist which confer certain characteristics on the scion, concerned with the scion's size and its resistance to soil-borne diseases.

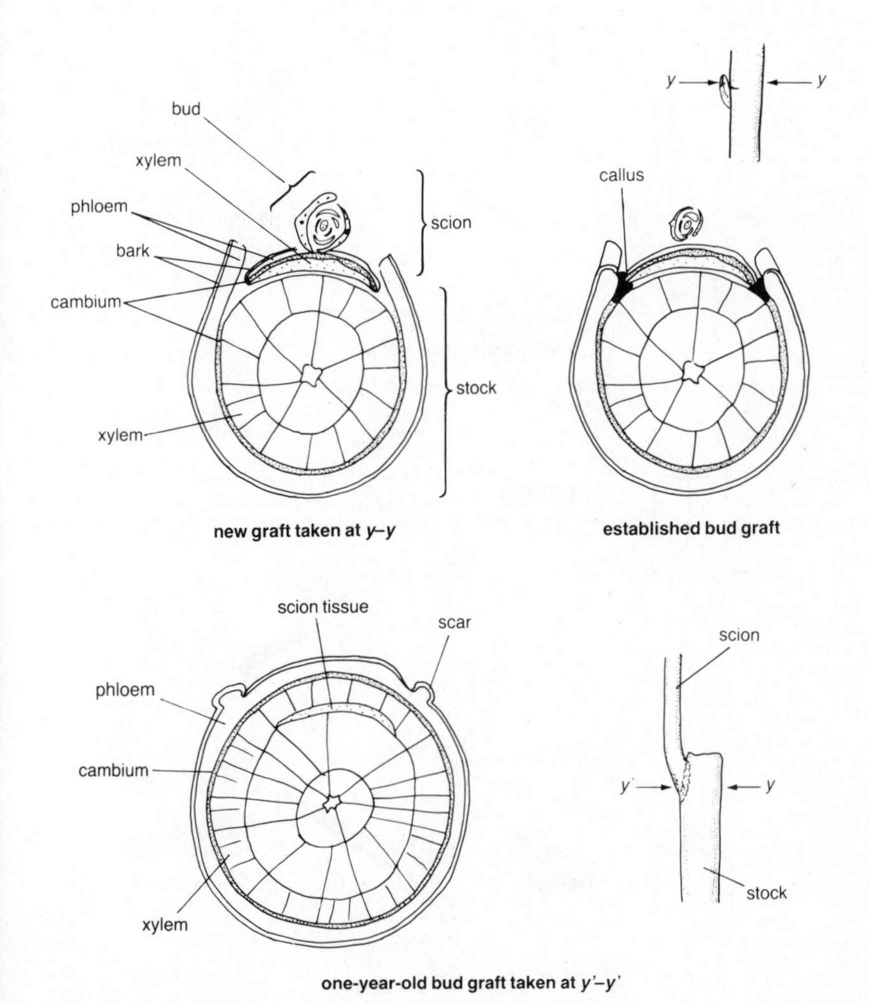

Figure 6.9 The development of tissue associated with a bud graft.

Research workers (at East Malling) have classified existing root-stocks for apple trees and bred new ones: the Malling series. The fruit from a particular variety is always the same because the tissue from which it develops (receptacle) is derived from the scion alone. Any variations in fruit quality which do exist – between plants, orchards or countries – must stem entirely from environmental (including nutritional) differences between the plants.

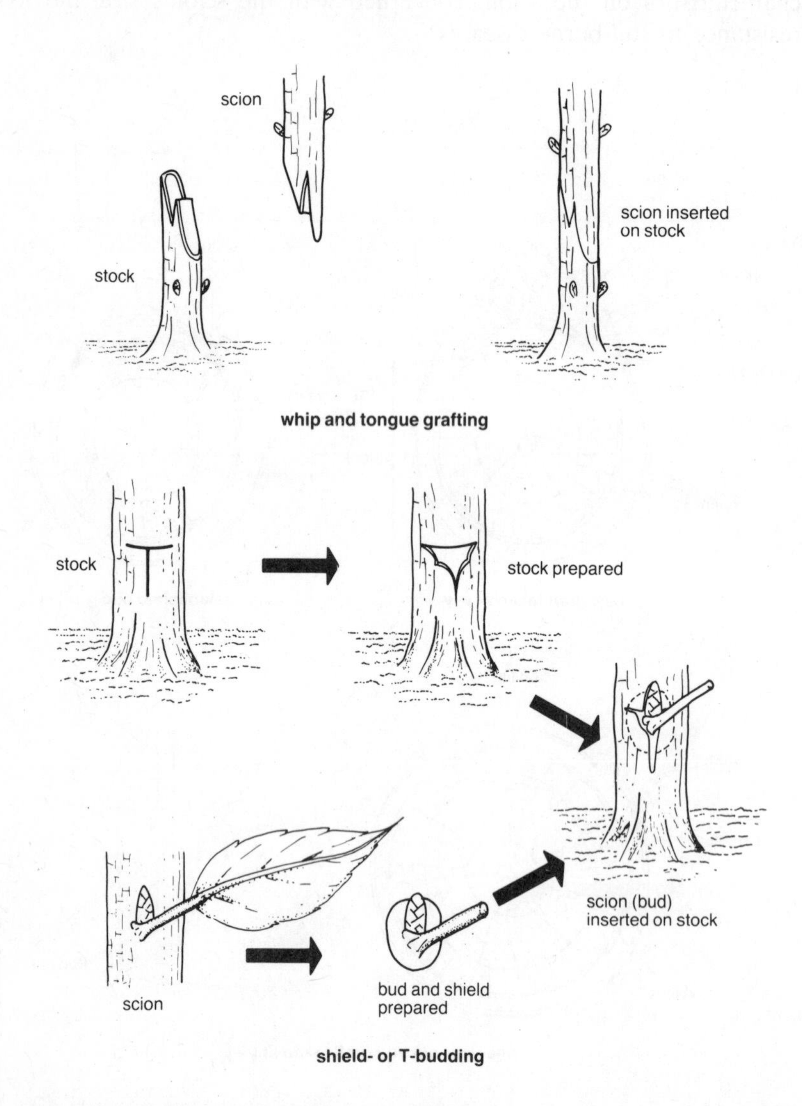

Figure 6.10 Whip and tongue grafting and budding.

Rapid propagation – meristem and tissue culture

While plant material may be made nutritionally independent and propagated by taking cuttings or grafting tissue onto root stocks, more artificial culturing is now possible. A very small portion of tissue, not contaminated with bacteria or fungi, is grown in the laboratory in an artificial environment which contains some energy-yielding substance (generally sugar) and appropriate nourishment. Because this tissue is cultured in an artificial feeding environment, we can control the *type* of cells formed by careful use of various plant hormones to which they respond. Under certain conditions, a whole host of single cells are produced, forming a khaki soup! When growing conditions change, the single cells behave just as would a single fertilised egg in an ovule. Each still derives nourishment and energy from the artificial environment, but develops into a small plant – this gradually responds to light, becomes autotrophic, and develops into a sexually mature plant. In addition to the environment being sterile, tissue (from, say, carrot phloem) is first grown in special thimble flasks which rotate, causing newly formed cells to slough off the original tissue. The medium in these flasks contains

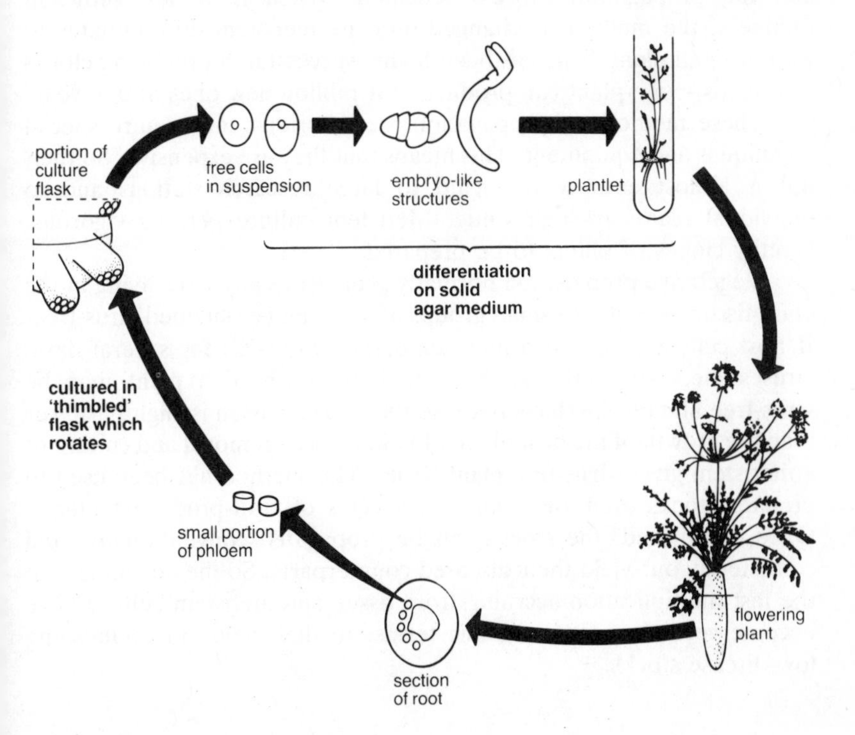

Figure 6.11 Vegetative propagation using tissue culture.

sugar, amino acids, minerals and specific concentrations of the plant hormones IAA and kinetin. By changing the ratio of IAA to kinetin and by putting them on a solid jelly-type medium (agar), the single cells are forced to divide and differentiate. (Figure 6.11 summarises these events.)

These techniques, still to be explored and perfected, give a method of reproduction rivalling that of the micro-organism. Plant hormones are used to trick the cells of a higher plant to behave as though they were single cells capable of producing only single cells; they are then allowed to develop according to their own genetic programme. Oil palm has been successfully propagated by these methods, which rapidly produced many uniform plants.

A similar method of propagation used to increase numbers very rapidly (to 'bulk up') is called **meristem culture**. A whole meristem, from the vegetative apex of a bud, is cultured in a sterile medium with artificial food. The small portion of tissue is differentiating and continues to do so in culture. It forms side 'buds' – yet more meristems – adjacent to the original meristem, and these structures can be removed and the propagation process repeated. When there are sufficient numbers, the medium is changed and the meristem differentiates to form a small plant. This has been highly successful in producing clones of orchids – one plant can produce four million new ones in one year.

These methods of propagation are laborious and require special techniques and equipment. This means that they are expensive. So their use is almost entirely restricted to large research stations and to individual plants of high value. Meristem culture permits vigorous, healthy clones of plants to be prepared.

Vegetative propagation normally propagates any virus living in the plant tissue as well, but small groups of cells can be obtained virus-free. If host cells are held at a high temperature (37 °C) for several days, viruses die. Alternatively, the actual meristem of a plant may be virus-free already (perhaps because the virus invasion is slightly slower than the growth of the host plant). In either case, removal and culture of a meristem gives **virus-free plant tissue**. This method has been used to prepare stocks of most temperate species of fruit-producing plants. Once established, the stocks can be propagated conventionally, and consistently out-yield their diseased counterparts. So the advantages of the fast multiplication accruing from tissue and meristem culture have become secondary, in economic terms, to their value in establishing low-disease stocks.

Summary

The production of surplus biomass from a breeding population is the basis of food production. We must assess the benefits of using mature or immature organisms for food:

a *mature* organisms are the only ones which can reproduce, and they tend to contain more food per individual than young ones;

b *immature* organisms produce biomass more quickly per unit of body mass than do older relatives, but we must be able to utilise this biomass as food, and the advantages of immature organisms must be sufficient to justify the maintenance of their parents.

The *rate* of reproduction often limits food production. Higher rates generally:

a will require fewer individuals to be maintained to breed replacements; but

b will require human effort.

7 The units of heredity – assembling the best

As we saw in Chapter 5, food production can be improved by changing the environment, but generally such increased output of food requires increased input. To increase output with the *same* input, we must improve the efficiency of the organisms themselves.

The characteristics of most food-producing organisms are largely determined by information stored in the nucleus of each cell. To be better able to improve existing strains, we must understand how this information is stored, and how it is gained by an individual from its parents. This is **heredity**, the topic of this chapter. We shall discuss practical breeding programmes in Chapters 8 and 9.

The science of **genetics** helps us to understand how differences arise among individuals. It tells us that this variation can be attributed to two causes: different inherited factors and different environments in which organisms grow. If we are to have the opportunity to choose the better organism for food production, variation must exist already. Knowledge of genetics enables us to assess the likelihood that any superiority shown by an individual will be passed on to its offspring.

Two facets of genetics are particularly relevant. The first part of this chapter discusses how the movement of chromosomes at meiosis effects the transmission of inherited characteristics. (It complements the specific description of the processes in basic biology texts.) The second part develops the principles of **variation**, the basis of improvement.

A basic knowledge of cell division and of the breeding experiments carried out by Mendel is assumed.

Heredity

Early experiments by Mendel

The mathematical analysis of the outcome of simple breeding experiments led Mendel to postulate that characteristics of an organism are

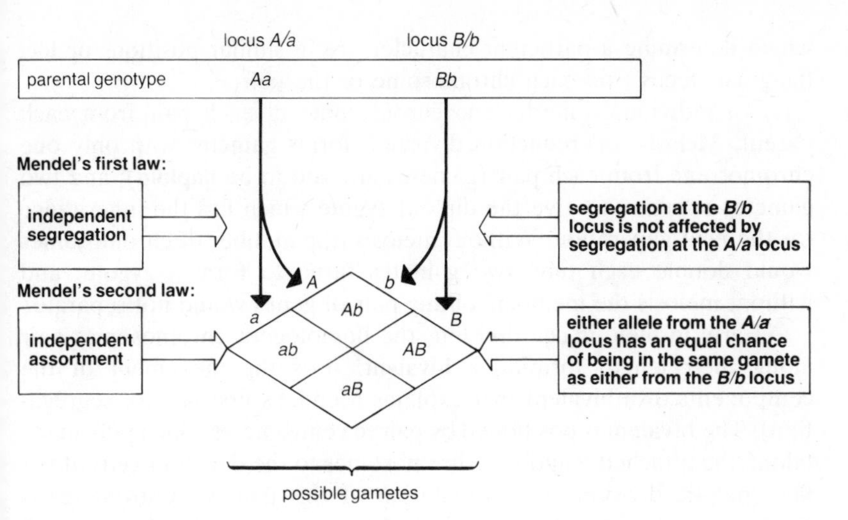

Figure 7.1 Mendel's laws.

controlled by **paired factors** or **alleles**. A gamete will carry one and only one of each pair (Mendel's first law, of **segregation**); and when a gamete is formed, each allele from any one pair may combine randomly with either allele from any other pair (Mendel's second law, of **independent assortment**). Figure 7.1 summarises these laws.

Mendel was careful to take true-breeding strains of peas. To be certain of this, he self-fertilised the original stocks for several generations until the offspring showed no variation, either among themselves or from the parent. He also carried out controlled pollination between known and intended parents.

He showed enormous perseverance in dealing with these test pollinations; the seeds were carefully planted, their characteristics noted and the number of each type recorded. This, alas, is the basis of much research; the careful, often tedious, collection of data to test a hypothesis. Nevertheless, he proposed the two laws of heredity in 1866. The scientific world was ill-prepared, and it was not until 1900, when the scientists of that day realised the implications of these laws, that the work was 'rediscovered'.

Inheritance and chromosomes

The body cells of most adult farm animals and plants have an even number of chromosomes; the chromosomes occur in so-called **homologous pairs**, and the individual is said to be **diploid**. The paired genes which determine a particular character are in similar positions or **loci** (singular 'locus') on each chromosome of the pair.

An individual inherits one chromosome of each pair from each parent. **Meiosis**, or 'reduction division', forms gametes with only one chromosome from each pair (gametes are said to be **haploid**); and two gametes combine to give the diploid zygote which has the inheritance for the next generation. Without meiosis, the number of chromosomes would double each time two gametes fused to form a zygote, and without meiosis the members of any pair of genes would not separate.

Early in meiotic cell division, the homologous chromosomes pair along their length forming a **bivalent**; it is the movement of the components of a bivalent that explains Mendel's first law (of segregation). The bivalent is positioned by paired centromeres which pull apart, taking the attached lengths of chromosomes to the daughter cells of the first meiotic division. The orientation of this pair of centromeres is random in relation to the direction of movement, so there is an equal likelihood of the copy of either chromosome of the pair going to any one of the gametes. The separation of the two members of a bivalent to different cells is the biological basis of what is described in Mendel's first law.

The movement of the members of the homologous pairs of chromosomes during gamete formation also can account for Mendel's second law (of independent assortment). The *random* orientation of the paired chromosomes during the early stages of meiosis leads to the equal likelihood that a given daughter cell will receive a copy of either chromosome of any homologous pair and hence of the genetic material that it carries.

So it can be seen that the inheritance patterns observed by Mendel last century can be explained in terms of the way that chromosomes move at meiosis; the 'chromosomal theory of inheritance', proposed early this century, effectively summarises these ideas.

Mendel's fortunate choice of character

While there are not normally any exceptions to Mendel's first law, Mendel's second law does not always hold true. It is broken when the loci controlling two separate characters are located on the same chromosome. The loci are said to be in the same **linkage group**, because the chromosome forms a physical link between them. If they are close together, there is strong linkage; if they are far apart, the linkage is weak.

The frequency with which genes at two linked loci segregate together depends upon the proportion of **crossing over** or **chiasma**

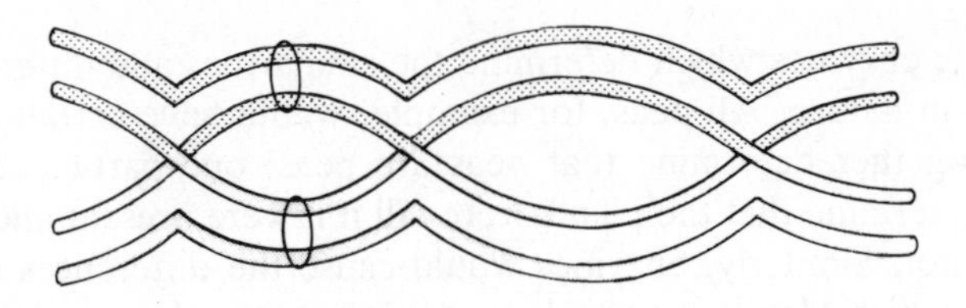

Figure 7.2 Chiasmata, during late prophase (diplotene) in cell division. The shading distinguishes the maternal from the paternal chromosome.

formation occurring in the bivalent. Formation of chiasmata (the plural of 'chiasma') results in genetic **recombination**, which provides the opportunity to produce new, heritable combinations of characteristics. Figure 7.2 shows chiasma structure at late prophase of meiosis, when the homologous chromosomes have each divided to form two chromatids after the earlier pairing. It should be emphasised that the daughter chromatids of each of the homologous chromosomes are paired on *both* sides of the chiasma. These crossovers are formed by breakage followed by wrong joining – careful thought should convince you that this is the only way that a chiasma can form, be successfully pulled apart as meiosis proceeds *and* result in genetic recombination.

Mendel was fortunate in his choice of characteristics in that they were on different chromosomes. He also happened to choose a gene at each locus with a **dominant** effect on the characteristic. Dominant genes exert the same effect in the diploid organism when in combination with either the same allele (the **homozygous** condition) or with any other allele at that locus (the **heterozygous** condition). The other alleles are by definition **recessive** to the dominant allele; there is no distinction in appearance between those individuals carrying one and those carrying two dominant alleles. The effect of the recessive is displayed only when homozygous. In Mendel's tall peas, the genetic constitution (**genotype**) at the height locus might have been either homozygous dominant (*TT*) as in the case of true-breeding plants or heterozygous (*Tt*) as in the case of hybrids, but those which were dwarf in appearance *must* have been homozygous recessive (*tt*).

Molecular expression of gene action

We often speak of alleles at specific loci being responsible for the characteristics of an organism, but just how is this possible? One of the first real attempts to explain gene action was the 'one gene, one enzyme' theory proposed by Beadle and Tatum in 1941. They argued that each gene was responsible for the production of a particular enzyme, and that it is these enzymes which determine the similarities and differences we observe in nature. All peas, for example, would have certain enzymes which together determine that peas are peas; one particular enzyme would determine that the plants were tall if it were present and dwarf if it were not. Similarly, enzymes would cause the differences in colour and size which Mendel's peas showed, along with other so-called **major gene** differences, such as the high lysine content of maize seed (Chapter 9) and genes which give plants specific resistance to disease (Chapter 10). Many modern textbooks will provide suitable information about gene action at cellular level.

In essence, the current concept of gene action is as follows. The double coil of individual molecules known as **deoxyribonucleic acid (DNA)** has four different component **bases**, which provide a code for the cell's activities. Considering bases in threes (triplets), there are 64 possible code units; theoretically, only 22 are needed to represent the 22 amino acids found in cellular proteins. The remainder provide duplicates and punctuation for the code. DNA is stored as a reference code in the nucleus and is replicated prior to cell division; its information is made available so that amino acids may be strung together in the correct order to make appropriate proteins.

The DNA master copy is utilised via a working code known as **messenger ribonucleic acid (mRNA)**. This also uses four distinct bases in triplets. Proteins are assembled on ribosomes, where **transfer ribonucleic acid (tRNA)** matches the mRNA triplet with the appropriate amino acid and this is joined, by means of a peptide bond, to the growing protein. Thus the sequence of triplets in the DNA determine the number and sequence of amino acids brought together to form the protein.

A protein formed in this way *may* act as an enzyme. An enzyme allows a reaction to take place more quickly, because it lowers the 'activation energy' of the reaction by bringing the reactants close together. If that chemical reaction allows, say, gibberellic acid to be formed in the plant, then the plant grows tall. If the enzyme is not formed, gibberellic acid is not synthesised, and the plant stays dwarf. Evidence of this nature led to the 'one gene, one enzyme' theory.

This model explains how a dominant gene masks a recessive gene in the heterozygote; the protein produced from the single dominant allele is sufficient for the reaction or event to occur, regardless of the non-functional abnormal protein synthesised by the recessive allele.

So we can see that because an individual inherits a set of chromosomes from each parent, there are two sets of instructions for each enzyme. If both alleles produce useless enzyme, the catalyst for a particular reaction is missing, and we may identify the resulting phenotype and describe it as that of a recessive homozygote. But if one allele can produce sufficient enzyme for the reaction, the aberrant allele will not be apparent – the heterozygote is phenotypically indistinguishable from the dominant homozygote.

Variation

The environment

The gross difference in height between the tall group and the dwarf group of Mendel's peas tends to mask the small variations which exist

Table 7.1 Distribution of heights of maize plants (after Emerson and East).

Height of plant (to the nearest 10 cm)	Range of height	Relative frequency
70	65–75	1
80	75–85	3
90	85–95	4
100	95–105	12
110	105–115	25
120	115–125	49
130	125–135	68
140	135–145	95
150	145–155	96
160	155–165	78
170	165–175	53
180	175–185	26
190	185–195	16
200	195–205	3
210	205–215	1

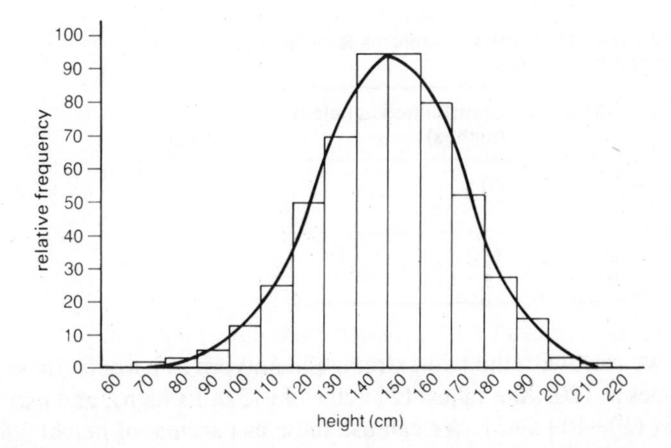

Figure 7.3 Histogram of height distribution in maize plants, and the curve which best fits the histogram.

among individuals within each group. Measurements of each individual in a population (Table 7.1) gives a frequency distribution as shown in Figure 7.3. The spread of this curve would increase if plants were grown in different fields on the same farm, and it would increase still further if they were grown in different counties or countries; the growing conditions in these various places would be quite dissimilar. This **environmental variation** results from the environment modifying the effects of a particular genotype to produce a range of phenotypes. Environmental variation has great significance for breeders; if superiority stemmed solely from favourable growing conditions, there would be no point in choosing a 'superior' individual from a population; this kind of superiority is not passed to offspring.

Small differences between individuals

As the environment can blur the categories into which we would put individuals, so genes may produce effects so small that we cannot recognise segregation at a particular locus. Variation is continuous in both cases because of this blurring effect, and gives the characteristic bell-shaped frequency distribution.

To appreciate how several genes lead to continuous variation, we can use a simple example. Suppose that two loci are involved in the determination of height: the '*A*/*a*' and the '*B*/*b*' loci. The loci are not on the same chromosome; the alleles at each locus act independently, and one allele adds to the effect of the other (gene action is additive). Let us also suppose that there is no environmental variation and that the contributions made to height are as shown in Table 7.2.

Table 7.2 The contributions to height made by the *A*/*a* locus and the *B*/*b* locus.

Locus	Allele	Contribution to height (inches)
A/*a*	*A*	20
	a	10
B/*b*	*B*	15
	b	5

If we start with the two extreme phenotypes as parents, these have genotypes *AABB*, the tallest (20+20+15+15 units high), and *aabb*, the shortest (10+10+5+5). We can use these as parents, of height 70 units and 30 units.

When these parents are cross-fertilised, the offspring – the F_1

population – have genotype *AaBb*, which gives a phenotype of 50 (20+10+15+5) units; the F_1 are genetically identical, and midway between the two parents. The F_2, obtained by intercrossing (or **selfing**) the F_1, has a range of genotypes from *AABB* to *aabb*; these are summarised in Table 7.3 and Figure 7.4.

The full implication of our model is now revealed. By plotting the proportion (relative number) of F_2 individuals found in each height range, we obtain the histogram (Figure 7.4) which is in basic outline a

Table 7.3 Variation in height between parents and second generation (after Lewis and John).

Type	Genotype	Relative frequency	Additive effect	Mean height (inches)
P_1	*AABB*	All	20+20+15+15	70
P_2	*aabb*	All	10+10+5+5	30
F_1	*AaBb*	All	20+10+15+5	50
F_2	*AABB*	1	20+20+15+15	70
	AABb	2	20+20+15+5	60
	AaBB	2	20+10+15+15	60
	AAbb	1	20+20+5+5	50
	AaBb	4	20+10+15+5	50
	aaBB	1	10+10+15+15	50
	Aabb	2	20+10+5+5	40
	aaBb	2	10+10+15+5	40
	aabb	1	10+10+5+5	30

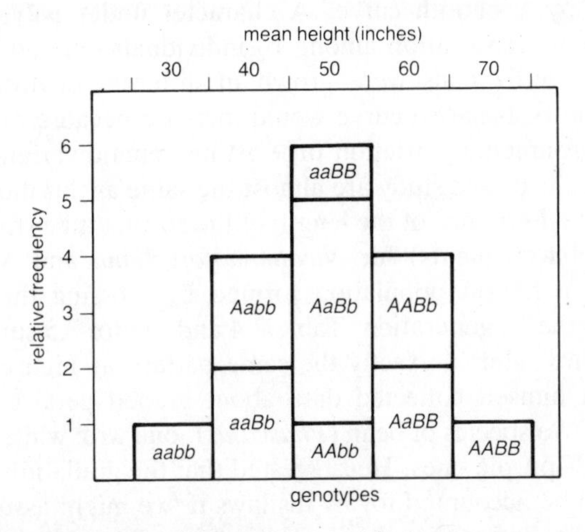

Figure 7.4 Mean height and relative frequency of F_2 types (from Table 7.3, after Lewis and John).

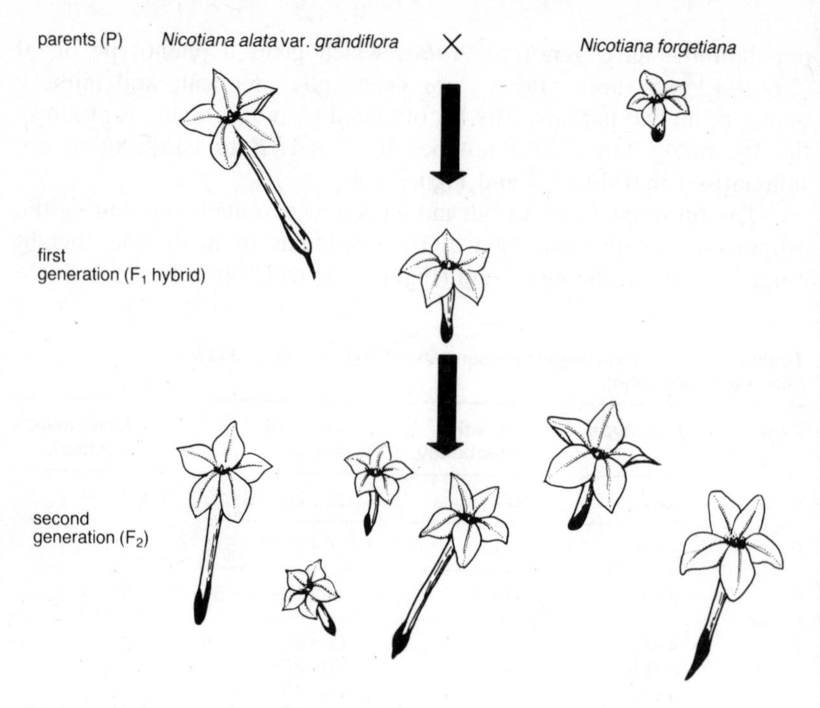

Figure 7.5 Variation in corolla length in tobacco plants (see Table 7.4).

bell-shaped curve, and all of *this* variation is of a genetic origin. In nature, there are usually more than two loci involved and the histogram is replaced by a smooth curve. A character under **polygenic control** shows continuous variation among F_2 individuals derived in this way.

If these individuals were grown in different environments, the spread of the bell-shaped curve would increase because of the imposition of environmental variation on existing genetic variation.

The results of one study are almost the same as this model. In 1913 East studied inheritance of the length of the corolla tube (formed by the petals of tobacco plants) for *Nicotiana forgetiana* and *N. alata* var. *grandiflora*, a hybrid population formed by crossing these two and members of the F_2 generation. Table 7.4 and Figure 7.5 summarise this work and show almost exactly the same pattern as Figure 7.4.

Mendel himself collected data about graded petal colour in the offspring of two species of bean (*Phaseolus*), one with white flowers and one with red–purple ones. He suggested that the gradation of colour in the F_2 could be accounted for by his laws if 'we might assume that the colours of the flowers ... is a combination of two or more entirely independent colours which individually, act like any other...'.

Table 7.4 Frequency distribution of corolla length in tobacco plants (after East).

Length of corolla (mm)	***N. forgetiana***	***N. alata* var. *grandiflora***	**F_1**	**F_2**
20	9			
25	133			5
30	28			27
35			3	79
40			30	136
45			58	125
50			20	132
55				102
60				105
65		1		64
70		9		30
75		50		15
80		56		6
85		32		2
90		9		
95				

So we can see that gradual variation in nature may be caused by a character being controlled by genes at more than one locus.

The way genes act

We normally think of Mendel using plant characteristics which were controlled by genes showing dominance, yet he did encounter and explain additive gene action with the inheritance of petal colour in *Phaseolus*. He probably encountered also a type of gene action known as 'over-dominance', but it is unlikely that he could explain this.

The true-breeding parental lines of peas which he used were 20–30 cm (short) and 180–210 cm (tall). From a particular cross, however, using parents which were 30 cm and 180 cm in height, all the offspring were between 180 cm and 230 cm tall. Tallness is **over-**

dominant. (This is important in hybrids formed between individuals of different origin; these tend to be more vigorous, showing what we call **hybrid vigour** or **heterosis**. We shall return to this in a later section.)

We can explain over-dominance by postulating that one gene can affect the expression of another gene at the same or at a different locus. An example illustrates this type of effect.

Bateson (who coined the word 'genetics') carried out experiments on sweet peas. Two pure-breeding, and therefore homozygous, white-coloured varieties produced purple-flowered offspring when crossed. The F_2 population, obtained by intercrossing the F_1 (selfing), had a ratio of nine coloured to seven white flowers. The ratio 9:7 seems to derive from the 9:3:3:1 ratio normally expected from a two-factor cross, so apparently two independent loci are involved. Individuals contributing to the '9' of the ratio have at least one dominant gene at both loci; those in the '7' may have a dominant gene at one or neither locus. For colour to be expressed, it seems, dominant alleles of both genes must be present.

Figure 7.6 shows in summary how alleles at one locus may interact.

The way in which gene action affects the work of plant and animal breeders is shown in Table 7.5.

The creation in practice of pure-breeding, inbred lines of organisms grown by us for food depends on the species tolerating self-fertilisation. Wheat normally undergoes self-fertilisation, but many organisms require cross-fertilisation. Lines (or breeds) of, say, cattle are not

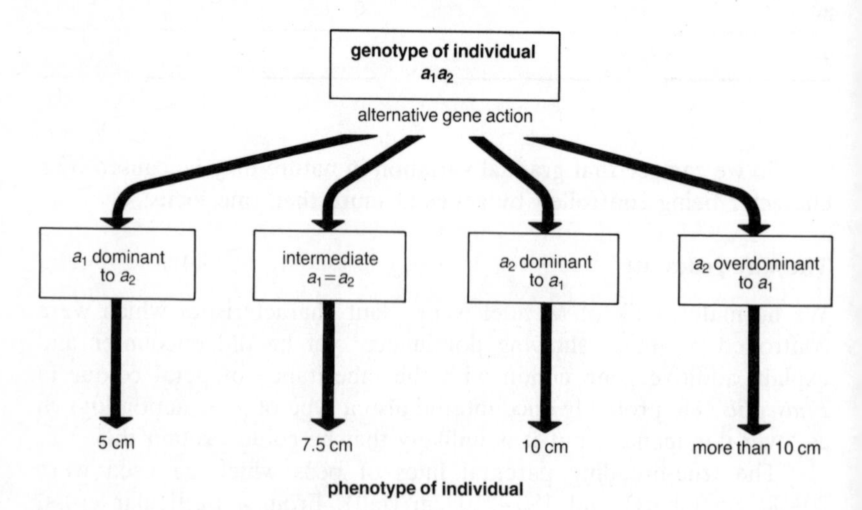

Figure 7.6 Interaction between alleles at *one* locus (a_1a_2). a_1 represents 5 cm of height and a_2 represents 10 cm of height.

Table 7.5 Gene action.

Between alleles at one locus	*additivity*	The best genotypes are homozygous for a particular locus and could be maintained as pure breeding lines.
	dominance	The best genotypes may be homozygous or heterozygous. The former will breed true while the latter will not.
	overdominance	The best genotype is heterozygous and so cannot be pure-bred. Heterosis is shown.
Between alleles at different loci	*gene interaction*	The best genotypes are likely to be cross-bred.
	additivity (polygenes)	The best genotypes are homozygous for a particular locus and could be maintained as pure breeding lines.

pure-breeding, but they are partially inbred; some loci are heterozygous and others are homozygous. Knowledge and understanding of these breeding systems dictates the breeding programme adopted for particular species, and we shall discuss them further in Chapter 8.

Vigorous growth by hybrids

Our own history has many examples of inbreeding which illustrate the long-term effects of close relatives marrying and having children – the frequent brother–sister marriages may have accounted for the general physical and mental decline of the Pharaohs of Egypt, and many societies have incest taboos well established in their religious laws.

Despite this realisation that inbreeding is beset with problems, many present-day varieties have been created by intense inbreeding. By the early 1780s, the Colling brothers had mated a bull called Favourite with his own offspring for five successive generations, so establishing the Shorthorn breed of cattle. Favourite himself was the result of close inbreeding of the offspring of a cow called Beautiful Lady Margaret.

A rather complex formula has been devised to calculate the genetic similarities between related organisms. A pioneer of these methods, Sewall Wright, investigated the relationships between individual Shorthorn animals in 1920. He calculated that had Favourite been alive then, the bull would have been more closely related to the average of all Shorthorns than a father is to his own son! At least 130 years after Favourite's birth, his genetic characteristics were largely similar to all Shorthorns. This herd proved popular and productive in the UK until the 1950s, when its dairy productivity was surpassed by that of the Friesian breed.

This account of the offspring of one bull argues against the inbreeding depression shown in Pharaohs; indeed, evidence from inbreeding depression and from hybrid vigour are often in conflict. The failure of the Shorthorn under present farming methods to compete with the Friesian may reflect ill-effects of early inbreeding. The mating of close relatives may have reduced genetic variation in the breed, variation that was necessary for it to be able to respond to opportunities of continued improvement.

Heterosis and inbreeding depression (its converse) are still difficult to explain genetically. We might expect that as long as gene action is either additive or dominant then inbreeding could take place without loss of vigour. When alleles at one locus interact, however, the heterozygous or outbred individual may be more vigorous than the inbred individual because of over-dominance.

Vigour may also derive from interaction between genes at different loci. When heterozygosity decreases during inbreeding there is just one allele at each homozygous locus so there is less likelihood of the genes interacting because there are fewer in the nucleus.

A further cause of inbreeding depression stems from using just a few individuals in a breeding programme. This may, quite by chance, perpetuate some non-useful genes. Other genes may be perpetuated because they are tightly linked to genes which the breeder values highly; if these linked genes have an adverse effect on vigour, the individual shows inbreeding depression.

There are now vast quantities of literature on the subject of heterosis; this is a highly simplified account. But Chapter 8 will show that this understanding is sufficient for us to breed hybrids; and it is clear that hybrids may out-yield their homozygous relatives – whatever the explanation.

Genes in populations

We have considered the way that genes within an organism are expressed, their effect on the phenotype, and how they cause individuals to differ. Breeders try to pick organisms with favourable genes to use as parents in a breeding programme. We now turn our attention to the factors which cause genes to become more or less common in a breeding population over several generations.

We would expect that without intervention the proportion of genes (their frequencies) would remain the same throughout many generations. G. W. Hardy and W. Weinberg, working independently, proposed this in 1908, and they listed the factors which cause gene frequencies to change. Changes occur when:

1 mating is not random between individuals;

2 there is mutation of the genes under consideration;

3 a small sample is taken which does not represent the whole population (genetic drift);

4 certain individuals are less successful than others so that the genes they carry are less likely to be passed on to the next generation (selection).

These conditions affecting gene frequency are summarised in the Hardy–Weinberg Law.

Considering this law for the frequencies of just two alleles at one locus, we can show that the frequency will remain constant, unless one of the above conditions operates. So, for the locus '*A*/*a*' where the frequency of *A* is 'p' and that of *a* is 'q', p and q will remain constant, and the proportion of the genotypes *AA*, *Aa* and *aa* will be $p^2{:}2pq{:}q^2$.

Mathematicians can introduce more than one locus, intensities of selection, known sampling errors, and known mutation rates, and then calculate the theoretical change in gene frequency after each generation. The computer lends itself well to the study of this type of model.

The Hardy–Weinberg law gives us the basis for breeding programmes, but the programmes themselves depend on *changes* in gene frequency and hence a disruption of the Hardy–Weinberg equilibrium. From the list on page 99, the breeder can select the best individuals for breeding, and cull the worst for immediate food consumption (1); he can introduce a new genetic base from different stocks often obtained from the centres of origin, and induce mutation by using ionising radiation (2); he can and must deal with small populations (3); and he will generally only allow certain selected matings (4).

Selecting the best genes

Assuming that we know what we wish to improve in our plants and animals (Chapter 9), we can select the best currently available and breed from it. This ancient practice domesticated and developed existing breeds (Chapter 4); modern breeders can be more decisive.

Selection takes one of three forms – stabilising, directional and disruptive (Figure 7.7) – all three may be used in artificial selection programmes.

Directional selection (Figure 7.7a) is by far the most important in genetic improvement. This type of selection aims to produce breeds and strains which are 'better' than the ones they replace. This usually means

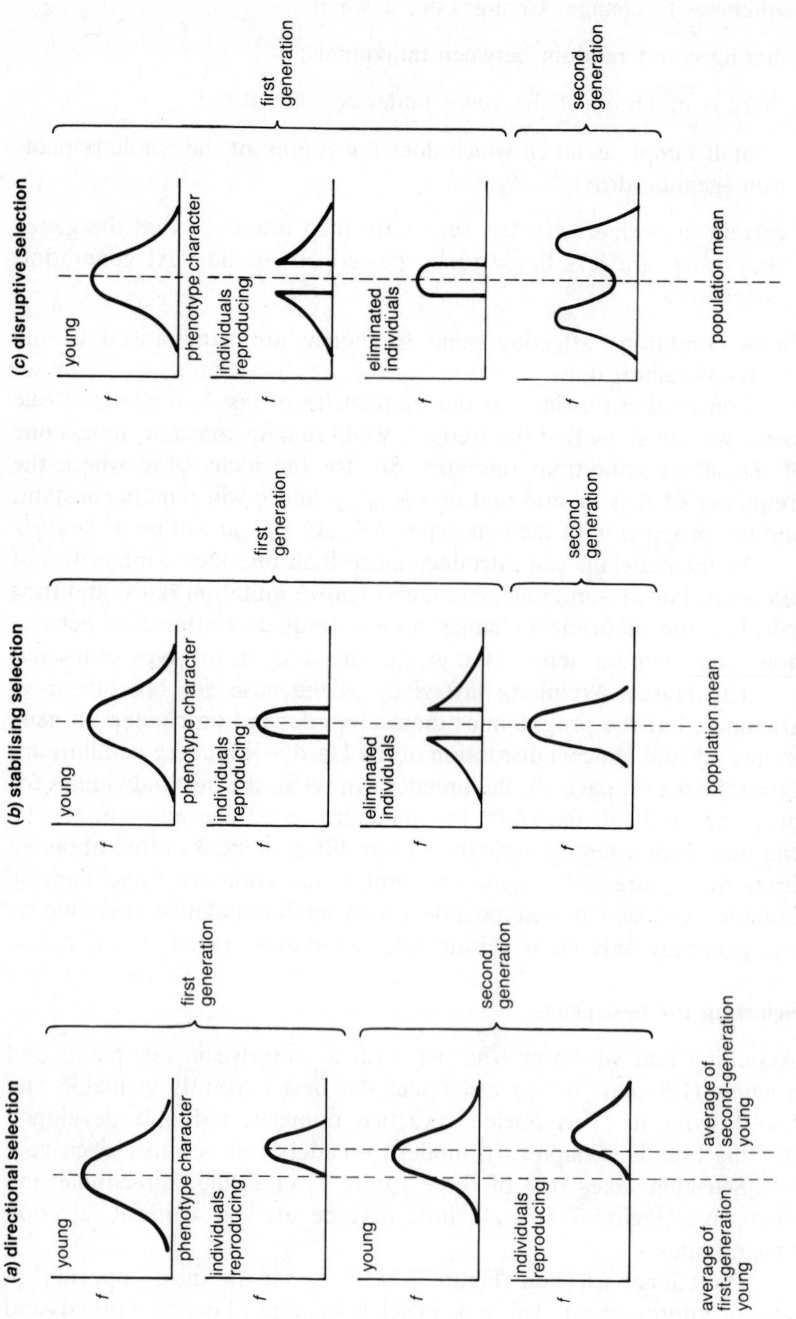

Figure 7.7 Types of selection (*f* is the frequency).

selecting individuals so as gradually to increase the frequency of those genes which contribute to the 'better' character. (Chapter 9 will discuss practical examples of directional selection.)

Stabilising selection (Figure 7.7b) is of particular interest to farmers, since they seek to grow plants and animals which have little variation; this allows the farmer to cultivate the organisms in a standard way.

In practice, both of these types of selection are applied together so that the population average (mean) shifts and the variability (spread about this mean) is reduced. (Note that characters such as 'horned' or 'hornless' in cattle, and requirement for vernalisation in winter varieties of wheat but not in spring ones, are really discontinuous traits, rather than the result of stabilising selection.)

Disruptive selection (Figure 7.7c) is employed when it is more convenient to select for extremes of a characteristic in one species to meet the requirements of different markets. Varieties of barley have been developed which are suitable for malting and subsequent beer-making and have low levels of nitrogen in the grain, contrasting strongly with the higher levels favoured for a feeding barley. Likewise, strong and weak wheats, used for bread- and biscuit-making, contain high and low levels of the protein gluten. Breeds of domestic animals have become more specialised, though true disruptive selection is not involved – Friesian cattle for milk, Hereford cattle for beef; Merino sheep for wool, Suffolk sheep for meat. Disruptive selection in the original population has become directional selection in the new sub-groups.

Passing on the best genes

We have considered the gene, its expression and the factors which alter its incidence in a population; we have seen how selection can be applied to a population to alter the mean and the spread of a characteristic. We now consider to what extent superior characteristics are passed from parents to their offspring; if a selected population does not pass on some of its adjudged superiority, there is little point in engaging in a selection programme.

Heritability is a measure of the proportion of variation shown by a population which has a genetic origin; it predicts the progress of a breeding programme. We can understand heritability by looking at Figure 7.8.

V_u is the mean of the unselected original population; V_s is the mean of the individuals selected from that population to be parents of the next. The difference between these means, the superiority of the selected parents, is S. V_o is the mean of the offspring population; and R

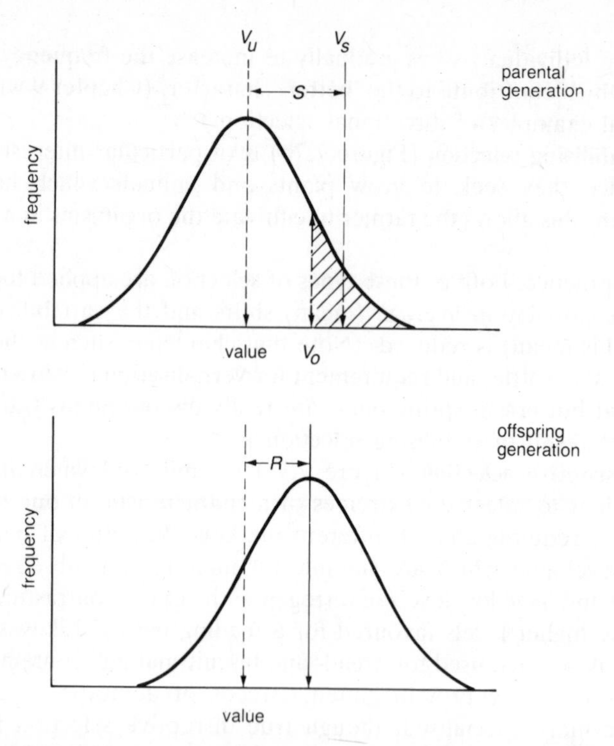

Figure 7.8 Distribution of parental, selected parental and offspring populations.

is the difference between V_o and V_u, the response to selection, when both parent and offspring generations are measured in the same environment. The parental population shows variation for the character of interest; a few extreme individuals are selected from this as parents (the shaded area). The offspring are all measured for the characteristic and their mean recorded.

The ratios of the values of R and S now give us a direct estimate of the proportion of the superiority of chosen parents which is passed on to their offspring. This is the heritability, and is usually written h^2. The equation $R = Sh^2$ enables us to predict the response to selection and assesses the usefulness of selection in an improvement programme.

Two examples may clarify the concept. Human eye colour is determined genetically and not by the environment. At the opposite extreme, phonetic ability has a very low heritability and regional variations in speech are totally dependent upon our environment, so $V_o = V_u$ and we can say that accent has a heritability of zero. (Our ability to *learn* a language will have a heritability above zero because this is affected by our genes.)

The heritability of a character helps us to decide whether a breeding programme will be successful and is therefore useful in the planning stages. The simplest way of estimating it is to plot the value for the offspring against that of their parents. The gradient of this graph gives the heritability, because we are measuring the extent to which genetically related individuals resemble each other. If the heritability is high, the resemblance is great, and the slope approaches one.

Other statistical methods may be used, but remember that we are *estimating* genetic variation, so we should not be surprised that there may be differences among the values obtained by different methods and in different populations. In addition, heritability may change as the activity and frequencies of the genes which it measures change during a selection programme; the genes of merit may become fixed and the population changed. In extreme cases we cannot improve further unless we improve the environment (Chapter 5) or introduce greater genetic diversity.

A word of warning

The environment can exert its effect through so-called **genotype × environment (G×E)** interaction. We saw that the growing of organisms on different farms or in different countries caused the population to vary round the mean of the population; some thrived, others weakened in their new different environments. Consider two breeds in one feeding environment. Let us assume that the mean of one population is greater than that of the other, so they can be ranked first and second. Imagine now that those breeds were bred and grown in a completely different feeding environment where the quantity and quality of nutrition is different.

We expect one of the three possibilities shown in Figure 7.9. If the

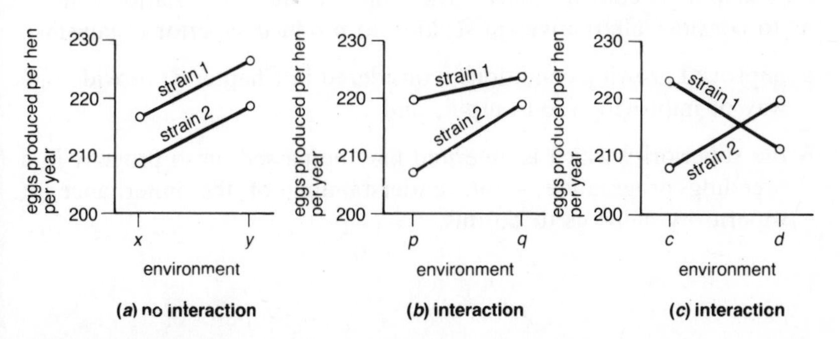

Figure 7.9 Genotype × environment interactions for poultry. (Based on a Ministry of Agriculture, Fisheries and Food publication.)

change is as in Figure 7.9a, no G×E interaction exists; both populations respond equally well (or badly) to the change in environment. The alternative responses both show that the strains respond *differently* in the changed environment. A G×E interaction exists when two (or more) strains differ in their performance relationships in different environments.

For example, breeds of cattle such as the Hereford, selected in temperate regions of Europe, may out-perform locally adapted breeds when introduced into tropical regions, but this only occurs when the Herefords are protected from local hazards. Without this protection, the well-adapted breeds such as the Afrikander or Brahman out-perform the imported Hereford. The reasons for such differences are not always obvious. For example, susceptibility to tropical ticks may lead to the low production of a Hereford, rather than its inability to withstand heat *per se*. By contrast, the high milk yields of Friesian cattle may be directly affected by heat.

The implications of these interactions are clear. We cannot select plants or animals as efficient food producers in one area, and then expect them to grow well in all other areas. We are often tempted to introduce high-yielding plants and animals already available into a region supporting low-yielding strains, but we should resist this temptation until these strains have been fully tested under local conditions. High-yielding strains may have to be bred locally and existing high-performance varieties may contribute significantly to these breeding programmes.

Summary

Variation exists in most populations and may be attributed to environmental and genetic differences. Knowing the cause of variation enables us to consider alternatives in seeking to produce superior organisms:

a improved growing conditions (considered in Chapter 5) provide one way of improving food output; and

b the superiority which is inherited may be passed on to progeny in a breeding programme – our understanding of the inheritance of superiority helps us to do this.

8 Natural and artificial breeding – an organism's choice of mate

Left to themselves, plants and animals may have very individual ways of choosing their mates. When close relatives (or even the one individual) provide both male and female gametes, **inbreeding** results; **outbreeding** requires the mating of unrelated organisms. Natural breeding systems can present the breeder with practical difficulties and sometimes with short cuts. We must understand how changing the established patterns of breeding, inbreeding, and hybrid vigour may affect the performance of plants and animals in an artificial selection programme.

Natural breeding systems

The range of natural breeding systems is wide, particularly in plants. At one extreme there is obligatory self-fertilisation; a flower, with the stamens removed, will never set seed. In the middle, there are plants which mainly self-fertilise but which *can* cross-fertilise, and those which mainly cross-fertilise but which *can* self-fertilise. At the other extreme – a few plants and all farm animals – there is obligatory cross-fertilisation, because one organism produces only one type of gamete.

Animals (outbreeders)

Farm animals are **dioecious** (single-sexed) and must therefore cross-fertilise. This encourages outbreeding, but, as we shall see, *our* choice of related parents can result in inbreeding.

Plants (inbreeders and outbreeders)

Usually plants are **monoecious** (eggs and pollen are produced by one plant). Some plants may be dioecious; *Asparagus*, for example, like the farm animal, must cross-fertilise, which again encourages outbreeding.

Some monoecious plants, such as maize, produce pollen and eggs in separate flowers; they are said to be **diclinous**. Male tassels and female ears form different single-sexed inflorescences (collections of flowers) which ensure pollination between *different* flowers and so force outbreeding.

In plants with **monoclinous** flowers (both eggs and pollen closely positioned within the one floral structure), either cross- or self-fertilisation can take place. Given species, however, tend to be either outbred or inbred. Wheat, for example, is an inbreeder; nearly all the seeds originate from self-fertilisation. By contrast, other species with monoclinous flowers avoid self-fertilisation in a variety of ways:

a **protandry** or **protogyny**, where the anthers and embryo sac mature at different times;

b **di-** or even **tri-morphic flowers**, where the spatial separation of the androecium and gynoecium is different in different plants, which encourages visiting insects to receive and deposit pollen on or from two or three different parts of its body;

c **incompatibility systems** which determine whether or not pollen will germinate on a stigmatic surface.

The control of compatibility is particularly interesting. In some plants, pollen will not germinate on the style of the same plant even though both are reproductively mature; such plants are said to be **self-incompatible**. A genetically different plant is required for reciprocal crossing. At least two varieties of apples are required in orchards for this reason; each pollinates the other, then both set fruit.

Compatibility may be controlled by specific genes inherited like any others, so we can account for self-incompatibility in genetical terms. We know that a gene might have not just two alleles at one locus, but several: a so-called **multiple allelic series**. A diploid organism will still contain only two alleles, but they will be two of several; a gamete will carry just one allele, but one of several, not one of two.

The alleles are usually identified as S_1, S_2, S_3 and so on, and their involvement in ensuring cross-fertilisation hinges upon one golden rule – like repels like. A pollen grain with, say, the S_2 allele will not germinate and grow on a diploid style with S_2 in its genotype, no matter what the other allele may be. It *will* germinate so long as S_2 is *not* included in the genotype of the style. Figure 8.1 illustrates the possibilities in this so-called **gametophytic incompatibility**.

The presence of incompatibility systems is important for two

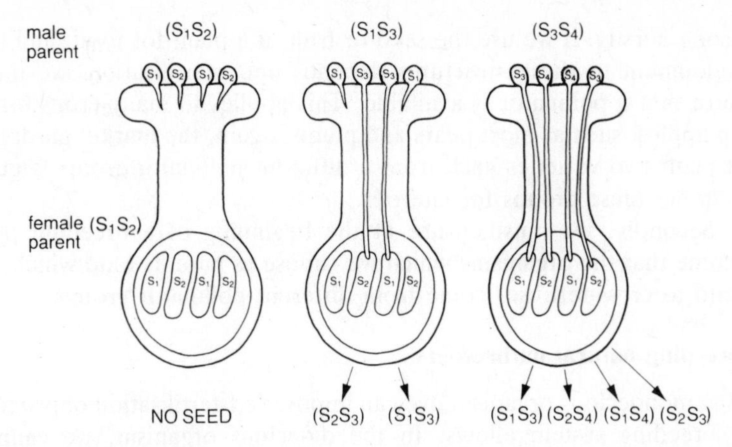

Figure 8.1 Styles and pollen of a plant with a multiple allelic series of incompatibility genes. Left: incompatible pollen. Centre: compatibility giving two genetic types. Right: compatibility giving four genetic types. (After Crane and Lawrence, 1947.)

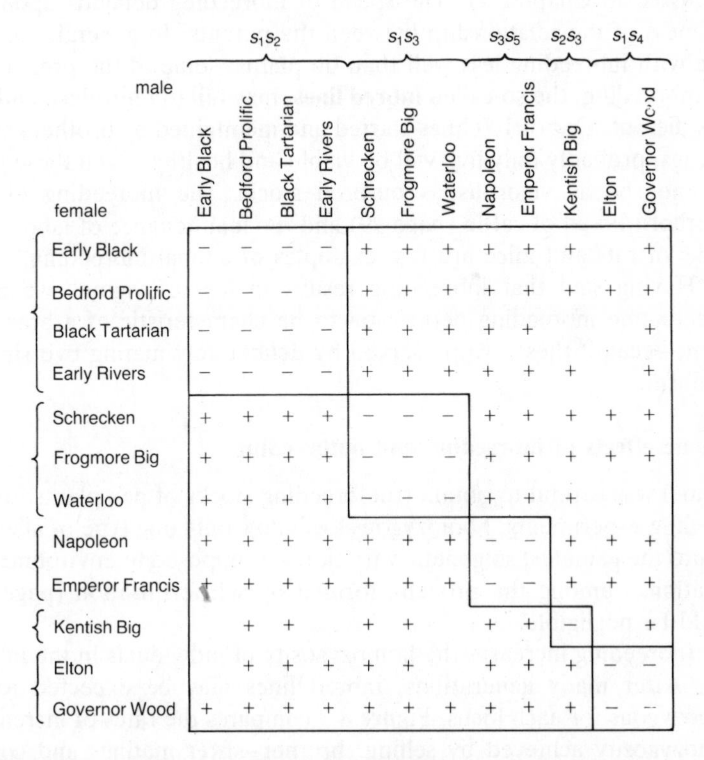

male →	S_1S_2				S_1S_3			S_3S_5		S_2S_3	S_1S_4	
female ↓	Early Black	Bedford Prolific	Black Tartarian	Early Rivers	Schrecken	Frogmore Big	Waterloo	Napoleon	Emperor Francis	Kentish Big	Elton	Governor Wood
Early Black	−	−	−	−	+	+	+	+	+	+		+
Bedford Prolific	−	−	−	−	+	+	+	+	+	+	+	+
Black Tartarian	−	−	−	−	+	+	+	+	+			+
Early Rivers	−	−	−	−	+	+	+	+	+	+		+
Schrecken	+	+	+	+	−	−	−	+	+	+	+	+
Frogmore Big	+	+	+	+	−	−	−	+	+	+	+	+
Waterloo	+	+	+	+	−	−	−	+	+	+		
Napoleon	+	+	+	+	+	+	+	−	−	+	+	+
Emperor Francis	+	+	+	+	+	+	+	−	−	+	+	+
Kentish Big		+	+		+	+		+	+	−		+
Elton		+	+	+	+	+		+	+	+	−	−
Governor Wood	+	+	+	+	+	+	+	+	+	+	−	−

Figure 8.2 Compatibility relations between twelve types of sweet cherry; there are five incompatibility groups. + indicates successful pollination; − indicates that pollination was unsuccessful. (After Crane and Lawrence, 1947.)

reasons. Firstly, if we use the seed or fruit of a plant for food, and the development of these structures depends upon fertilisation, we must ensure that a pollinator is available. This applies to many crops other than apples, such as most pears and plums; again, the market gardener will plant two varieties each from a different pollinator group. Figure 8.2 shows these groups for cherries.

Secondly, we must ensure at the beginning of a breeding programme that the organisms which we choose as parents (and which we intend to cross-fertilise) come from different pollinator groups.

Inbreeding natural outbreeders

In the monoecious organism, we can impose self-fertilisation only when the breeding system allows. In the dioecious organism, we cannot impose self-fertilisation, but we *can* inbreed the strain by choosing closely related individuals as parents. Inbreeding results in inbreeding depression and its converse, hybrid vigour, stems from outbreeding (discussed in Chapter 7). The speed of inbreeding depends upon the closeness of the relationship between the parents. In general, animals cope with inbreeding less well than do plants; some of the progeny of this inbreeding, the so-called **inbred lines**, may fail to reproduce and the lines die out. Out of 100 lines started and maintained by brother–sister matings, probably only five will be viable and healthy. Even these lines may not be as vigorous as outbred stock. The inbreeding of the Shorthorn breed of cattle (page 99) and the maintenance of laboratory stocks of rats and mice are two examples of animal inbreeding.

Having said that inbreeding results in loss of vigour, we must expect some inbreeding depression to be characteristic of a breed or strain, because these are preserved by deliberately mating two similar organisms.

Genetic effects of inbreeding and outbreeding

Mendel was careful to obtain true-breeding stocks of pea plants for his breeding experiments; homozygous loci allow only one type of allele to go into the gametes, so genetic variation – as opposed to environmental variation – among the progeny formed by self-fertilisation (page 93) would be negligible.

Inbreeding increases the homozygosity of individuals in the inbred line. After many generations, inbred lines may be expected to be homozygous for each locus. Figure 8.3 compares the rates of increasing homozygosity achieved by selfing, brother–sister matings and cousin matings.

By contrast, outbreeding tends to maintain heterozygosity at the

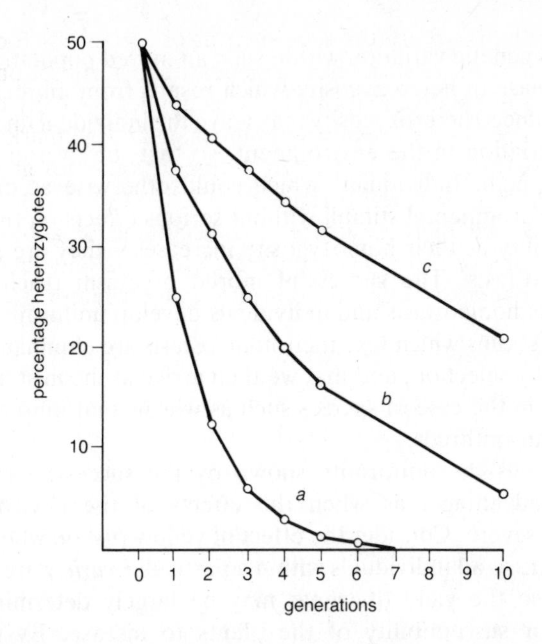

Figure 8.3 The decreasing percentage of heterozygotes in successive generations, following (*a*) selfing; (*b*) brother–sister mating; (*c*) first cousin mating. (After Lewis and John.)

loci, and hence in individuals and in the population as a whole. The degree of heterozygosity might then affect the variation (see page 98).

All individuals derived from one organism by asexual means have the same genotype, so the only source of variation is the environment; hence the actual phenotypic variation among individuals is less than that in a comparable outbred population (Table 8.1).

It is difficult to predict whether a population derived by inbreeding from one organism will show more or less variation than one obtained by outbreeding; hence the question mark in Table 8.1. It is certain that

Table 8.1 Variation among individuals in an identified reproducing population, which we call a breed, strain or variety.

Breeding system	Heterozygosity in individual	Origin of variation	Degree of variation within a strain or variety
sexual–outbreeder	high	G + E	high
sexual–inbreeder	low	E	low?
asexual	conserved	E	low

there is less genetic variation within such an inbred population, because of the decrease in heterozygosity which results from inbreeding. However, the reduced heterozygosity may leave the individual unable to cope with the variation in the environment, so that the overall phenotypic variation is high. Individuals, which could otherwise accommodate a range of environmental stimuli without serious effects on their growth, lose this ability as their homozygosity increases – they are said to lose their homeostasis. The successful inbred organism (such as wheat) maintains its homeostasis and individuals develop uniformly. It may be that inbred strains which lose their homeostasis are eliminated from the population by selection, and that we then arrive at the mistaken general conclusion, in the case of species such as wheat, that inbreds are more uniform than outbreds.

Sometimes the uniformity shown by the successful inbreeder is itself a disadvantage, as when the effects of the environment are particularly severe. Consider the effect of yellow rust on wheat; wheat is an inbreeder, so all individuals within a particular *variety* are genetically alike, and so the yield of wheat may be largely determined by the resistance or susceptibility of the plants to disease. By contrast, a *heterogenous* population of wheat plants might be less susceptible to yellow rust; while some plants might succumb and die, others would survive, and at least some harvest would be made. The yield of the heterogenous population, however, might be well below that of the inbred population in the absence of disease, since it would have low- as well as high-producing plants. (We shall return to disease in Chapter 10.)

Starting a breeding programme

Two types of breeding programme are possible. One type leads to a uniform population made up either of crossbreds derived from two or more inbred populations (which have the advantage of hybrid vigour) or of inbred individuals themselves. When F_1 hybrids are used for food, new individuals must be continually 'recreated' from the original parents. The second type relies upon variation within outbred populations, either variation in long-term outbreds, or variation released when hybrids reproduce sexually. This variation permits us to select individuals of 'superior' merit and to allow them to reproduce in preference to 'inferior' organisms. In this second type of programme, selection and restricted breeding continue through several generations to give gradual improvement in the value of the population. In both, though, we must ensure that the chosen parents actually mate.

Ensuring that the chosen parents mate

With higher animals we may ensure that a particular female is not fertilised by other than the chosen male simply by penning the two animals together or by the use of artificial insemination. Because we physically prevent any other male from mating, the pedigree of offspring produced from an ensuing pregnancy is known. For the plant breeder, on the other hand, the problems of ensuring that chosen parents cross-fertilise are often considerable. It is difficult to monitor the transport of pollen; the pollen may be contaminated as it leaves the flower of the outbreeder or may not effectively enter the flower of the inbreeder. In fact, we follow Mendel's example and use controlled pollination.

When self-fertilisation is known to be unlikely in outbreeders, simply enclosing both parents in a pollen-proof bag is sufficient to control pollination; foreign pollen is excluded.

To ensure a cross between two naturally inbreeding plants is more difficult. Perhaps the most effective way is physically to remove the anthers from the flower bud before it matures or they make contact with the stigma. This is called **emasculation** and can readily be carried out with needle and forceps. When the gynoecium of the emasculated flower is mature, pollen from the other chosen parent is placed on the stigma using a small paint brush. Again, no stray or foreign pollen must be allowed to make contact with the stigma prior to pollination and fertilisation, so the emasculated flowers are protected in pollen-proof bags.

A word of warning. Plants may not breed as one might expect from the flower structure, so we must know the pollination mechanism before we start on a breeding programme. Take, for example, the peas used by Mendel. We think of legumes (the plant group which includes peas) as being typical cross-pollinated flowers, with large showy petals and the style enclosed in the sheath-like keel ready, it appears, to protrude at the right moment and make contact with pollen carried on a visiting insect. Not a bit of it! Mendel chose this plant because it was an inbreeder. Apparently, the pea developed structures which best fitted it to cross-pollination, but later changed to inbreeding.

Hybrids

We have now brought together the genetic characteristics of our chosen parents.

Inbred parents tend to be homozygous at many loci, so the hybrids they form (the F_1) show genetic uniformity; any phenotypic variation stems from the environment. To produce the next generation (the F_2, obtained by interbreeding or selfing), heterozygous loci in the F_1 undergo genetic recombination (page 91); the genetic uniformity is lost and the variability among the F_2s will be greater than that among the F_1s.

By contrast, many of the loci in *outbred* parents are heterozygous; parents vary within their lines and F_1 populations are generally heterogeneous. Relative to the overall variation within a species, however, the F_1s tend to be uniform and this is favoured by modern farming techniques. And we know that hybrid vigour (heterosis) is shown by the hybrid, so that – other things being equal – the hybrid will outyield its inbred counterpart. Fortunately hybrids can readily be formed in plants and in animals, so we can use them in modern agriculture.

Hybrids of natural outbreeders

Generally some degree of inbreeding has been imposed upon the outbreeding plant; as we saw above, controlled pollination is an easy process. This reduces the variation in the population and allows ready characterisation of strains which can then be used as parents. One example of hybrid formation is in maize, perhaps the plant hybrid most widely grown from seed. This is an outbreeder and has diclinous flowers. In practice, inbred lines are established by repeated selection among the offspring of related plants. The potential of these inbred lines as parents in hybrid production is then tested; strain A is crossed with B, C, D and E, strain B is crossed with A, C, D and E, and so on. The merit of resulting hybrids is evaluated. The 'combining ability' of several potential parents is assessed prior to commercial hybrid production.

The commercial maize production procedure is quite simple. A row of strain A is grown alongside a row of strain B, one strain having the male tassel removed so that it is forced to be a female parent. Progeny obtained by planting the seed grow with the expected vigour, but the yield (number) of seeds is low because the female parent strain is inbred. To overcome this, a double hybrid is produced, a 'hybrid between two hybrids'. The two hybrid parents show vigour and in turn this female parent produces high yields of seed for the farmer (Figure 8.4). This double method of seed production replaced the F_1 hybrid technique in 1918, so the production and use of hybrid seed was well under way soon after the birth of genetics.

Animal hybrids

Like the outbred plant, the farm animal readily cross-fertilises to form a hybrid. However, the production of hybrids for commercial use depends

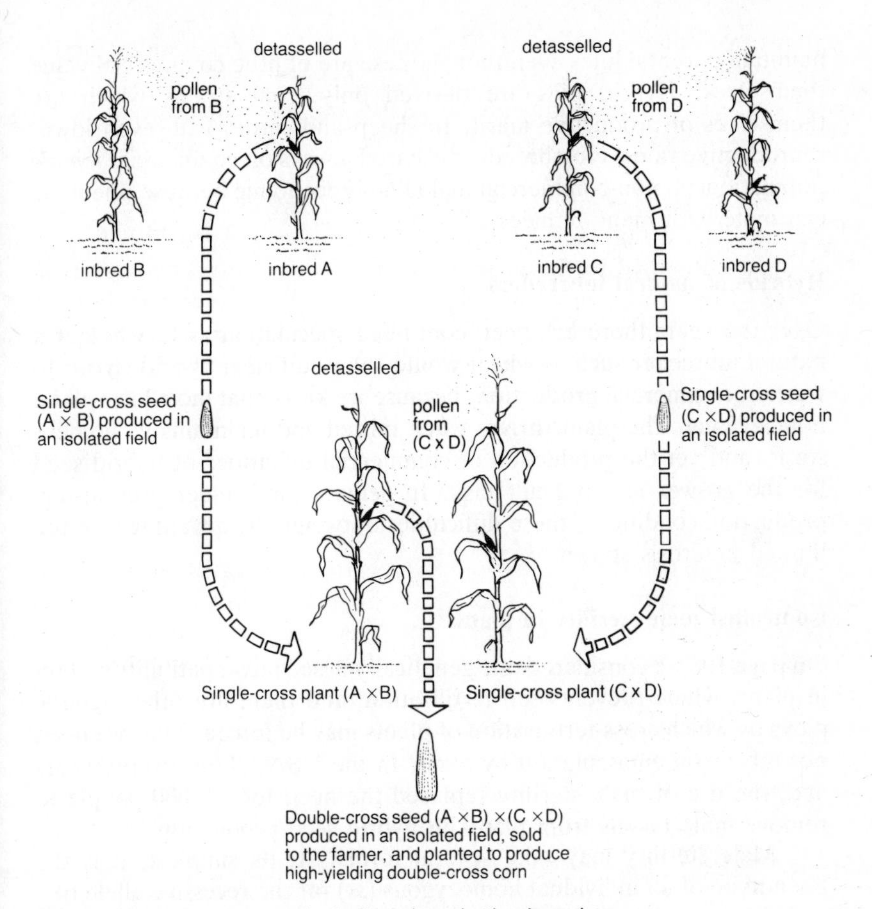

Figure 8.4 Double-cross method of seed production in maize.

upon a favourable balance between the merit (or extra value) of the hybrid, and the problems (or cost) of maintaining parental lines. The benefits of rearing hybrids for food must outweigh the drawbacks of being forced to keep, as parents, populations of animals which do not contribute directly to efficient food production.

These parental lines, as with maize hybrid production, will be slightly inbred; we know this because we distinguish one breed from another. Parental lines have lower reproduction rates than the hybrids which they give rise to, and are generally less vigorous. The economics of maintaining parental lines solely to produce hybrids are determined by the rate of reproduction of the species as a whole. The slower this is, the more animals are needed as parents to replace those used for food (Chapter 6). Poultry have high rates of reproduction, so we can maintain parental lines even though these are of little commercial value themselves. Hybrid pigs are derived only from strains which are themselves of productive merit. In sheep and cattle, with even lower reproductive rates, crossbreeding is based always on commercial *female* populations; a non-commercial male line is economic because one male can mate with many females.

Hybrids of natural inbreeders

Over the years there has been continued speculation as to whether a natural inbreeder such as wheat would show sufficient hybrid vigour to warrant commercial production, because we know that most loci will be homozygous. The plant thrives when inbred and maintains its homeostasis, and yet the production of commercial quantities of hybrid seed for the grower is a difficult task. In general, and under present-day production conditions, these difficulties outweigh the advantages of the limited heterosis shown.

Controlled male sterility in plants

On page 108 we considered one genetically based incompatibility system in plants which prevents self-fertilisation, and there are other genetic ploys by which cross-fertilisation of plants may be forced. Thus we need not rely upon emasculation by hand. In the USA, about twenty years ago, the use of male sterility replaced the need for 100 000 people to remove male tassels from plants in hybrid seed production.

Male sterility may take several forms. At its simplest, it is the phenotype of an individual homozygous (*ss*) for the recessive allele of a particular gene. Pollen (*s*) formed by this male-sterile individual cannot grow in the style of the same plant, so self-fertilisation is prevented, but the plant can act as a female parent to other pollen. By planting this and the intended male parent side by side, the cross can be made. This type of male-sterility can be perpetuated by crossing the male-sterile individual with a heterozygous individual (*Ss*); heterozygous and homozygous recessive individuals are produced, and the half of the progeny which will be self-fertile must be removed as soon as they can be identified. (Identification can be aided by linking the recessive male-sterile gene to a recessive one for, say, hairy leaves.)

Some types of male sterility are inherited via the cytoplasm and this means that they are transmitted maternally. The sterility arises from self-replicating factors in the cytoplasm which make the plant produce non-functional pollen. The egg contains more cytoplasm than the pollen grain, so that all of the offspring of the female parent, with the male-sterile factors in their cytoplasm, will be male-sterile. The hybrid cannot produce seed unless this male sterility can be removed, and genes are known which do just that – they restore normal fertility. Sugar beet, maize, flax and other crops may show this type of male sterility; we employ this labour-saving device to ease our manipulation of genes for production in the organism.

Reproduction of the hybrid

By controlled reproduction, desired features of two parents have been brought together in the hybrid offspring. What is the reproductive fate of this hybrid?

As we have seen, we may decide that its sole purpose is to be used for food. If its formation is easy, its production features excellent and its crop vigorous and uniform, we may be prepared to form the hybrids repeatedly; this is true of maize, for example. The drawbacks of the increased variability of the progeny (F_2) of the F_1 hybrids may be such as to outweigh any difficulties of producing more F_1 hybrids.

The reproductive future of hybrids may be limited biologically, however. Some hybrids are themselves infertile, because their parents are so different. Without our biological intervention they have no reproductive future. But other hybrids may give rise in the F_2 to variation which we can use for further genetic improvement. We shall consider each of these fates.

Hybrid infertility

To see why hybrids are often infertile, we must consider what happens when we try to mate dissimilar individuals. If they are very dissimilar, they will not mate; they lack similar mating behaviour and so they do not recognise each other sexually. If they do mate but rarely, the egg may be fertilised, but the embryo usually dies; the nuclear and cytoplasmic contents of the gametes are so different that normal development is prevented. There are exceptions where the hybrid develops normally to adulthood. We exploit some of these; for example, the mule is a hybrid from a donkey (*Equus asinus*) and horse (*Equus caballus*), with the horse as stallion. The mule, though, is infertile.

In the plant kingdom, apparently illegitimate matings are more frequent. Species in the same genus or even in different genera may breed together freely. The more distantly related the parents, the less likely is the hybrid to be fertile. We can intervene in two ways to ensure a reproductive future for infertile plant hybrids.

Firstly, the hybrid can be propagated by vegetative means; these we discussed in Chapter 6. These techniques are generally reserved for species where the individual plant is of high value (as with roses and apples), or where the method of vegetative propagation is simple (as with potatoes).

Our second method is the formation of **allopolyploids**. Infertility in hybrids can be explained when we consider chromosome movement during the production of gametes. Normally, homologous chromosomes pair during the prophase of meiosis; they then replicate to produce pairs of sister chromatids, held together at the chiasmata. The four chromatids go into the four cells produced by meiosis. But in the hybrid, the chromosomes are of different genetic origin and so cannot pair at prophase in meiosis; the resulting disarray makes the hybrid sterile.

If their chromosome number is doubled, hybrids may reproduce by sexual methods; each chromosome now has a homologous partner for the meiotic division and the plant *behaves* as a diploid. Certain drugs double the chromosome number and so create an allopolyploid by inhibiting the separation of chromosomes at mitosis. One commonly used drug is colchicine, which is obtained from the meadow saffron, *Colchicum autumnale*.

These laboratory manipulations make fertile the infertile hybrid. This enables the hybrid to reproduce easily and gives us the opportunity to select for further favourable characteristics in this newly created sexually reproducing population.

Plant breeders have employed this technique to combine the hardiness of rye (*Secale*) with the yield and grain quality of wheat (*Triticum*). The resulting *Triticosecale* has made a great impact on agriculture. We saw in Chapter 4 that wheat itself is thought to have evolved by hybridisation and doubling of chromosome number, but we must remember that this occurred quite naturally by chance. Coffee, sugar cane, oats, plum and tobacco are other crops which have evolved as allopolyploids.

Selection patterns

We return now to the original cross between selected parents which creates the variation from which we select new strains. As we saw, there will be genetic variation among F_1 individuals only if at least one parent comes from a population showing genetic variation. If both parents are largely homozygous, the F_1 will be genetically uniform, and we must wait until the F_2 for the massive release of genetic variation. We can now select chosen characteristics among individuals of subsequent generations; several options are available.

Culling and bulk selection

The simplest way to improve the genetic merit of a strain is to use only 'superior' organisms as parents. The converse of this is to remove individuals which we consider to be of little merit.

In animal improvement we call this **culling**. The animals are prevented from contributing genes to the next generation. Because of their high value and the slow rate of reproduction, these culled animals are used for food. This type of selection took place during the domestication of our present breeds.

Only in nearly natural or in the most extensive methods of production from wild or semi-wild animal populations do we use culling solely. Generally, mating programmes are arranged specifically to introduce new genes into the herd or to concentrate favourable ones. Because one male can fertilise many females, the introduction of superior males is a frequent method of herd improvement. So it was that in the latter part of the 1780s the Shorthorn bull Comet, an offspring of Favourite (page 99), sold for 1000 guineas!

Often, different herds within one breed are of similar merit and specific mating programmes allow genes to be introduced from one herd to another; the whole breed improves slowly. For a single herd to be superior to others in the breed, the farmer must plan his mating and culling programme very carefully.

The effect of introducing a superior male can be increased by use of **artificial insemination (AI)**. Table 8.2 shows the amazing numerical potential of AI in the introduction of new genes to a large breeding population. For cattle, a further advantage of AI is that sperm can be stored for a long time. This means that we can actually check the effects of the genes – by recording the progress and productivity of a small group of the bull's progeny – before releasing the bulk of the sperm for insemination (page 136). In animal improvement, culling is often associated with the introduction of new genes by the purchase of superior individuals, or sperm, from other breeders.

Table 8.2 Data on artificial insemination in farm animals.

Species	Volume of semen obtainable from a single service (cm^3)	Easily frozen	Number of inseminations from a single service
horse	50–100	no	90
cattle	5	yes	500
sheep	1	no	25
pig	200–250	no	35

In the plant breeding programme, by contrast, the individual of low merit can not only be prevented from breeding, it can be discarded, having low individual value as food. This, and the much higher reproductive rate for plants, allows the selection in a bulk selection programme to be much more rigorous than in the corresponding animal programme.

In the initial stages of a bulk selection programme, the criteria used for selecting superior plants may be very coarse. An example will illustrate this. A programme was devised to recreate potatoes from their South American ancestors (*Solanum andigina*). The selection of superior plants was achieved by growing the whole population in the south of the UK where potato blight infection (see Chapter 10) is very likely to take place naturally; in addition, spores of the disease were sprayed on susceptible strains which were grown close by. This ensured that plants which did not show signs of potato blight were resistant. Another coarse (but efficient) way in which research workers selected useful plants was to collect only large tubers, by using a large mesh riddle on the mechanical harvester which separated the potatoes from the soil. These two simple techniques gave an initial selection which eliminated disease susceptibility and small tuber formation. More specialised, and often laboratory-based, assessments then took place to ensure that the potential commercial plant conformed to the needs of the market. Remember that we can *fix* certain characteristics of a strain of potato because it is propagated vegetatively. Our present-day potato, *Solanum tuberosum*, is an outbreeder; if we were to propagate this by seed rather than by tuber, variation among the sexually formed offspring would be high.

Bulk selection in a cereal may be equally coarse over the first few generations. We might apply known disease agents and harvest later than usual to ensure that disease resistance and grain retention are selected. More precise aspects of yield are selected for in later generations, in the knowledge and with the benefits of the initial bulk selection.

Pedigree selection

The pedigree of a plant or animal is the history of its parentage. Carefully chosen parents are used for the initial cross in pedigree selection; progress and merit of all offspring are recorded thereafter. Individuals are individually selected; those which show low merit are removed. Individuals are chosen which perform well. Matings are between specific individuals, not merely at random between a group of superior males and a group of superior females. Each subsequent generation is assessed with equal care. So, in contrast to bulk selection, where broad criteria are used to select a *group* of individuals as parents of the next generation, pedigree selection assesses *each* individual in *each* generation.

Changes in heterozygosity during selection

After the initial cross at the start of our breeding programme, the process of sexual reproduction further assorts the genetic material, as discussed on pages 94–100. Independent assortment of alleles during meiosis has different implications for the inbreeder and outbreeder, and often highlights the difference between plants and large animals.

For the inbreeder (plants only), there is a rapid increase in homozygosity. After as few as nine generations, a cereal may be registered with enough accuracy to preserve the breeder's royalty rights. When the new strain is released, it will breed true because, as we saw on page 00, the strain is largely homozygous.

By contrast, the mating of the outbreeder must be carefully controlled to minimise the rate of increase in homozygosity (the rate of inbreeding). Indeed, as much care is needed as was exercised over the parents in the initial cross. If the deliberate aim were to inbreed the strains and so realise the potential of a hybrid formed when two such lines are crossed, as in maize production, we would need to ensure that the rate of inbreeding was much less than that tolerated by normally inbreeding plants.

So, for the outbreeder, only general qualities may be attributed to the new strain; the strain is best considered as a population with generally superior merit; its characteristics cannot be embodied in one individual.

Backcrossing

Mendel used a backcross to establish the unknown genotype of an individual by crossing it with a known homozygous recessive genotype, a **test cross**. The ratio in the offspring of individuals in different classes revealed the gametic frequency in the unknown.

The breeder uses this in a slightly different way, to associate just one characteristic of variety A with the sound properties of yield and quality found in variety B. It is often used where variety A is a wild strain with a worthless yield and yet possesses, say, resistance to a particular disease, while variety B is susceptible to the disease so that its high yield potential is at risk. In combination, disease resistance and high yield could create a strain of superior merit. An example in animals

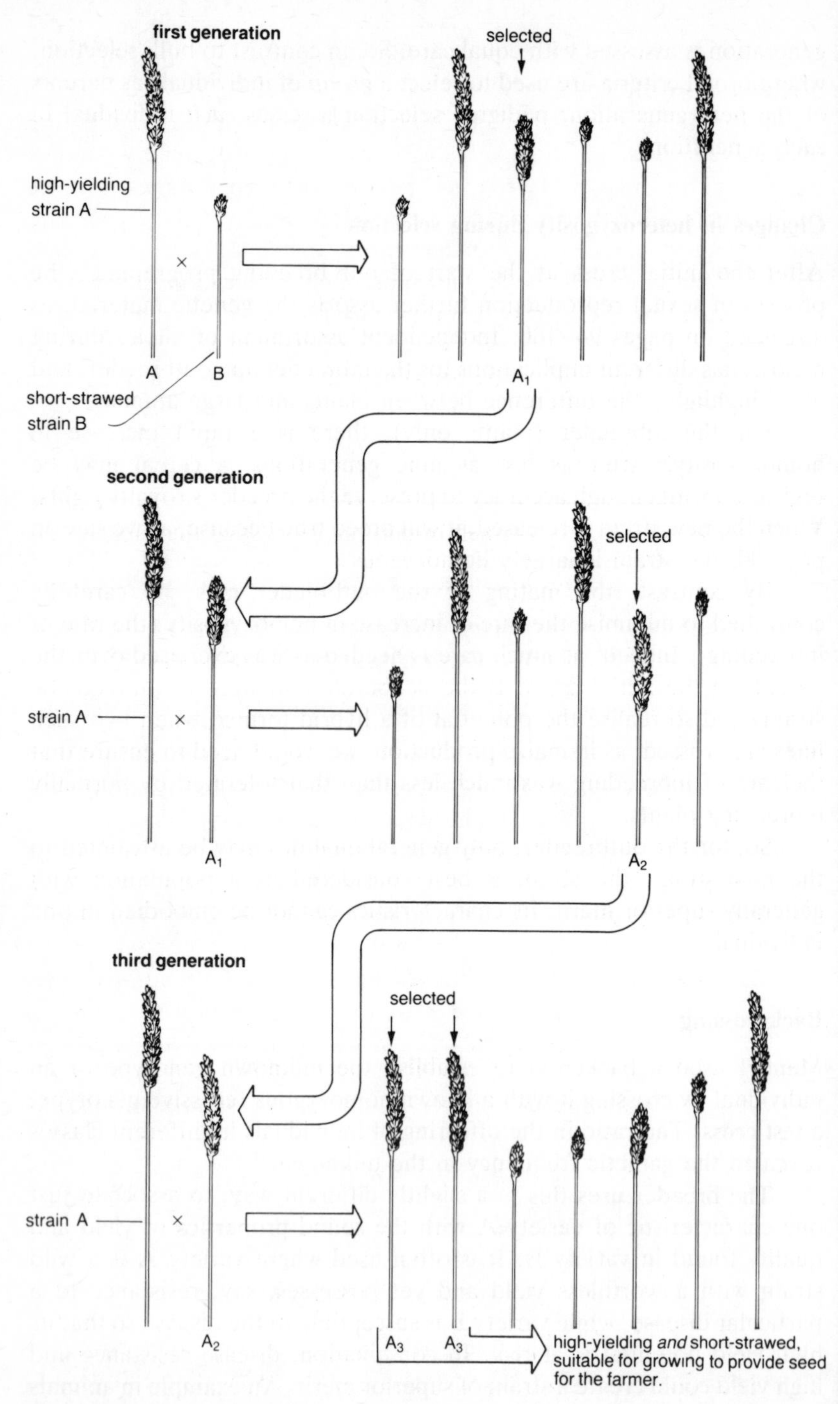

Figure 8.5 Theoretical scheme for backcrossing: strain A is high-yielding; strain B is short-strawed but low-yielding.

would be to introduce the 'polled' (hornless) gene into an otherwise desirable horned breed.

It is possible that an EEC regulation will soon require the skinning of sheep heads at the time of slaughter, to improve slaughter house hygiene. This may add impetus to breeding programmes to poll Blackface sheep; polled heads can be skinned economically by machine rather than by hand. The breeder sets up an initial cross and then follows a strict regime; the F_1 and subsequent generations are always crossed with parent B, but only if they have the single desirable feature of parent A (Figure 8.5). After several generations of backcrossing, the single feature of parent A has been incorporated into variety B.

Summary

There are several ways in which organisms reproduce themselves:

a asexual reproduction, or cloning, results in a uniform population;

b sexual reproduction between individuals, from dissimilar but uniform populations, produces a *uniform* hybrid population;

c sexual reproduction between a variety of *dissimilar* individuals gives a variable population.

The *food producer* favours a population of organisms which are uniform because he treats them all in the same way; hence breeds, strains and varieties with particular characteristics have been developed within species.

The plant or animal *breeder* requires inherited variation in order to develop the superior organism for food production.

These requirements are met by appropriate use of the patterns of reproduction, so we find that:

methods **a** and **b** propagate superior organisms, but require human manipulation;

method **c** is the basis for the creation of the superior organisms and requires direction by humans.

9 From genetic theory to breeding practice – making the decision

The yield of food from an organism has first to be identified. The quantity and quality of food derived from the organism, and the point in the growth of the organism at which it is harvested, are all important. Having identified yield, we shall consider actual examples where breeding programmes have altered the organism's yield. Considerations are not only biological; organisms are grown on a large scale for economic reasons, and the law of supply and demand operates. Governmental control of marketing may directly affect the food producer.

What do we mean by 'yield'?

Partitioning

We rarely use the whole of an organism for food: part is discarded. One way to increase yield, therefore, is to increase the proportion in the organism of high-value food; but we must remember that some food residues *are* of value in our society. Offal and hide are processed. Cereal straw is used as animal bedding, as fuel in specially designed straw-burning stoves, and increasingly (after treatment with alkali to improve its digestibility) in ruminant diets. For sugar beet, the leaves can be fed to cattle (after the poisonous oxalic acid has been broken down naturally by wilting) or dug into the soil where they act as a valuable green manure (page 27). The pulp remaining after extraction of sugar is also a valuable and attractive cattle food; it contains twenty per cent sugar, and also cell wall material which ruminants can digest more easily than the fibre found in cereals. There is some truth in the saying that we use every part of a pig except the 'squeak'!

The division of biomass into 'yield' and 'residues' is sometimes called **partition**. Modern varieties of wheat have more grain in the total biomass, and similar improvements have been made in maize, potatoes and many other crops. Plant breeding has made progress primarily by altering the partition and improving the so-called harvest index rather than the total production of biomass. With animals we favour a high 'killing-out percentage', whereby the carcass weight forms a high proportion of that of the live animal. Certain characteristics of the carcass are also important, such as the proportions of bone and fat; for several years we have preferred pig breeds with a low-fat carcass, and now lean meat in cattle and sheep is growing in popularity.

The quality of yield

Often the selling of food rests upon visual appeal – purchasers may buy food on the most irrational grounds. Therefore fruit and vegetables must be free of blemishes left by disease, though the buyer does not see the chemicals used to kill the disease agent! And certain breeds of animal have flesh which is more highly coloured when jointed, a trait favoured by the butcher.

We now consider aspects of yield in several of our food-producing organisms, aspects which illustrate the way we investigate, understand and then act upon features of food quality.

When the demand for lean pig meat became great, pig rearers sought breeds which had little fat on the carcass; the Pietrain is such a breed (Figure 9.1). Unfortunately, the meat is pale and so does not look very attractive to the buyer, and it oozes fluid before sale, so is not favoured by the butcher – the joints are losing weight and hence money. To resolve this conflict, two lines of investigation were pursued, one by the food technologist to find out why the meat was different, the other by the physiologist to see what feature of the live pig resulted in these unacceptable changes.

The food technologists, looking at chemical changes in the meat postmortem, found that the meat became pale because the muscle became too acid too quickly. The acid made cell membranes leaky, which accounted also for the loss of fluid from the tissue. The fall in pH postmortem is due to a build-up of lactic acid. Because oxygen is no longer supplied to the muscle, it becomes anaerobic; the whole tissue respires anaerobically via glycolysis and produces lactic acid. (Remember that muscle contains glycogen which provides the glucose for the breakdown pathway.)

The rapid production of acid is explained by the physiology of the pig while alive. The metabolic rate (MR) of the Pietrain pig is elevated

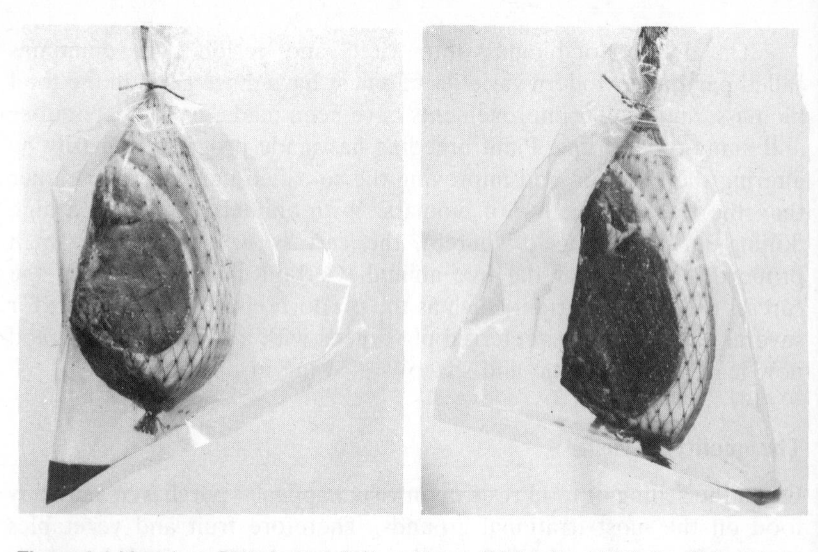

Figure 9.1 Meat from Pietrain pigs (left) and Large White pigs (right); the Pietrain has a low-fat carcass which exudes tissue fluid. (Courtesy of the Agricultural Research Council Meat Research Institute.)

just before slaughter because the pig suffers from stress due to the changed environment. Other pigs, such as the Large White, do not have this sudden increase in MR; the tissues break down the stored glycogen much more slowly, so the fall in pH is gradual. (In Chapter 11 we shall see that a low pH in food, including meat, helps to preserve it by preventing bacterial growth. Indeed, this natural postmortem effect can be maximised by feeding sugar to the pigs during the twelve or so hours before slaughter; the store of glycogen is increased, so more lactic acid is formed when the glycogen is broken down, and this helps preserve the meat. The ideal is a slow fall to a low pH.)

The Pietrain, then, has unusual sensitivity to stress. We can try to identify its inheritance. Many differences between the Pietrain and other pigs are attributable to a single recessive gene (or perhaps a small, tightly linked group of genes). The stress-sensitive homozygous recessive individuals can now be identified without slaughtering them, which is very useful if we are to preserve pigs without this drawback as potential parents in a selection programme.

The stress-sensitive pig (or human for that matter) responds in a particular way to halothane, an anaesthetic. This drug impedes the breakdown of lactic acid; the animal stiffens in what is known as the 'malignant hyperthermia syndrome' (MHS). So we can identify stress-sensitive animals at, say, eight weeks, before any decision has been made about their breeding future.

The occurrence of pale, soft exudative (PSE) muscle varies among breeds; in Pietrains nearly 100 per cent of individuals show it, whereas only eleven per cent of British Landrace and three per cent of Large White show it. Its incidence in humans, in the form of MHS, is one in 10 000.

Only recently has so much attention been paid by the large growers, packers and retailers to the visual appeal of food. Their interests are clearly those of increasing sales. However, it is almost impossible to measure the *flavour* of food by chemical methods. Twenty years ago, when techniques such as gas–liquid chromatography became available cheaply, it became easy to separate chemicals and hopes ran high that flavour could be measured and so quality controlled. Yet we still rely upon a tasting panel to assess flavour, as we rely upon words to describe that flavour. It *is* possible to measure *texture* in food, using a tenderometer, which assesses the mechanical toughness of the material. This device is quite straightforward when used with peas, but it is interesting to note that the mechanical properties of cotton require complex instrumentation to measure accurately the length, fineness, strength and uniformity of fibres.

It may be that the quantity of a particular *chemical* in a food varies in relation to the food's quality; if so, this presents a simple test. In foods for animals, we try to relate the food use to some chemical index. It is easy to measure the amounts of fibre, nitrogen (and hence protein), soluble carbohydrates and essential amino acids. The digestibility of a food can be estimated by feeding trials, or by placing it in contact with rumen liquor (obtained from an animal via a permanent tube, or cannula).

Sometimes the relationship between the quality and the quantity of a chemical is more complex. For example, the importance of protein content in determining quality in barley depends on the way in which the barley is to be used. The bulk of it is fed to farm animals, which may gain from it most of their nitrogen. For beer-brewing, however, the barley must have a low protein content or the beer will be cloudy (page 163). Barley used in the manufacture of grain whisky must have a high protein content because enzymes in the barley are used to break down the starch in other cereals before fermentation. We can identify the protein content by converting the nitrogen in it to ammonia (inorganic nitrogen) and titrating this against an acid. A simple conversion factor (6.5) scales the weight of inorganic nitrogen to represent the weight of protein.

Rather complex chemicals may be important components of yield, as are the enzymes used in grain whisky-making, the amylases. Their action is measured by recording the digestion of starch in starch jelly to which a small portion of the germinated grain has been applied. Iodine/potassium iodide solution forms an intense blue–black colour when starch is still present, whereas it gives no colour with the products of starch breakdown. We can even identify the particular enzymes responsible for this 'diastatic power' (DP) by means of electrophoresis, in which an electric field separates individual proteins because the proteins are themselves charged. The proteins which have DP are then indicated by adding starch to them and treating the result with iodine solution, the method by which one would identify 'spots' on a paper chromatogram.

Yield comprising more than one characteristic

In the majority of improvement programmes, we are concerned with several features of yield. Consider the simple example of the yield of sugar beet per hectare. The total yield of sugar Y is the quantity of roots sent to the refinery R multiplied by the sugar content of each root S: thus $Y = R \times S$. In seeking to improve total yield, it may be simpler to select for a *component* of yield, because each component has a higher heritability (page 103). Thus we can select for *total* sugar yield in one of three ways.

One method is **tandem selection**, in which selection is for one component of yield (say, tonnes of roots/hectare) at the start of the programme, and when this has reached the anticipated improvement, selection is made for the other component (sugar content of roots). While one component is under selection, the other is totally ignored. This method is much less efficient than simultaneous selection for both characters and is rarely used in practice, but it must have figured largely in the subconscious process of domestication discussed in Chapter 4.

A second method, in which selection is for two components at the same time, is to use **independent culling levels**. The breeder sets certain acceptable levels of performance for each trait; individuals falling below these levels for *either* component are rejected irrespective of their performance in the other component. Figure 9.2 applies this to the selection of poultry for broiler (meat) production. For our example, both root yield and sugar content must show superiority for the individual to merit inclusion in the next stage of the breeding programme.

Breeders nowadays tend to use a **selection index**, the third method. Each yield component is given its own particular weighting; the magnitude of this weighting reflects both the value of an improvement in that component and the expected ease of the underlying genetic change. Thus both inheritance (heritability and genetic correlations, page 103)

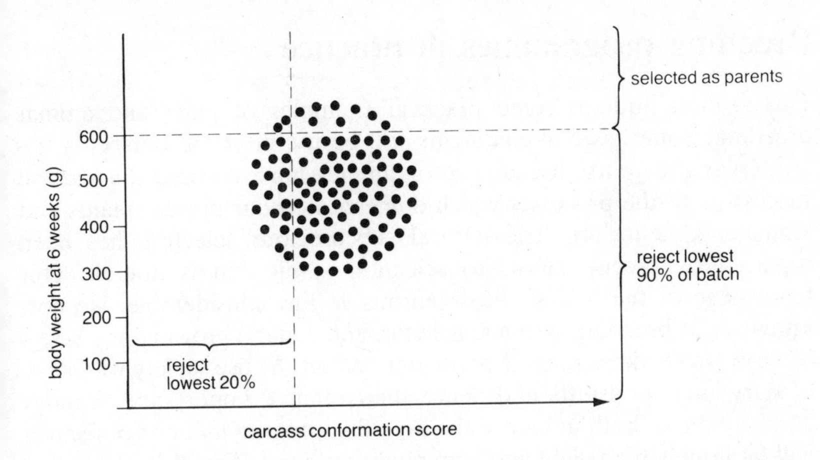

Figure 9.2 Independent culling levels operating in the selection of breeding stock for broiler (meat) production in poultry. The 'carcass conformation score' is high for birds with sound legs (the animal must be able to stand) and for birds without a sharp keel (a well-developed sternum leads to blistering of the skin).

and economic weightings affect selection indices.

The merit of sugar beet is now given as $M = aR + bS$, where a and b indicate the weighting given to the two components – the larger the weighting factor, the greater the importance of that component. M is the economic merit rather than the biological yield Y.

Practical aspects of selection

How does selection take place in practice? We should try to offset theoretical efficiency (which we would like to be high) against the desired ease of assessing the merit of the organism when grown on a farm.

An illustration may clarify this. Suppose that sheep are to be selected – as potential parents in a breeding programme – to combine fecundity (ability to reproduce), good growth and good wool quality. Each animal could be assessed using a selection index: its features are recorded and reference made to the fecundity of its own parents; the index is computed in the laboratory, the animals are revisited, and the actual selection made. By contrast, the ease of selection by independent culling levels may be preferred, so that (for instance) an animal must be born a twin, weigh over 20 kg at weaning and have wool finer than 25 μm. Animals with these features would automatically be chosen as parents. This selection could take place on the farm and remain simple under a range of conditions.

Breeding programmes in practice

This section outlines some practical examples of plant and animal breeding. Some general comments will be useful at the outset.

Over the years, breeding programmes have become the natural successors to the processes which domesticated our present plants and animals (Chapter 4). The natural subconscious selection has been replaced by precise conscious selection – but that is not all. Our knowledge of the biology of organisms is now considerable; we now know about breeding systems, genetics and general growing processes. We can also make economic predictions about the use society will make of particular foodstuffs and hence their relative importance. Biology and economics both influence the breeding programme; our objectives will be largely high yield and high quality of food. The objectives must be matched with the biology of the species, exploiting variation and selecting for chosen characteristics (as outlined in Chapter 8). This process may be summarised as in Figures 9.3, 9.4 and 9.5.

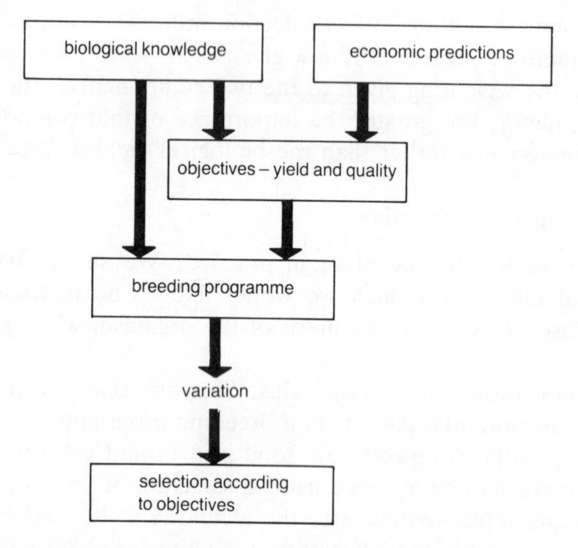

Figure 9.3 The principles behind the practice of plant and animal breeding.

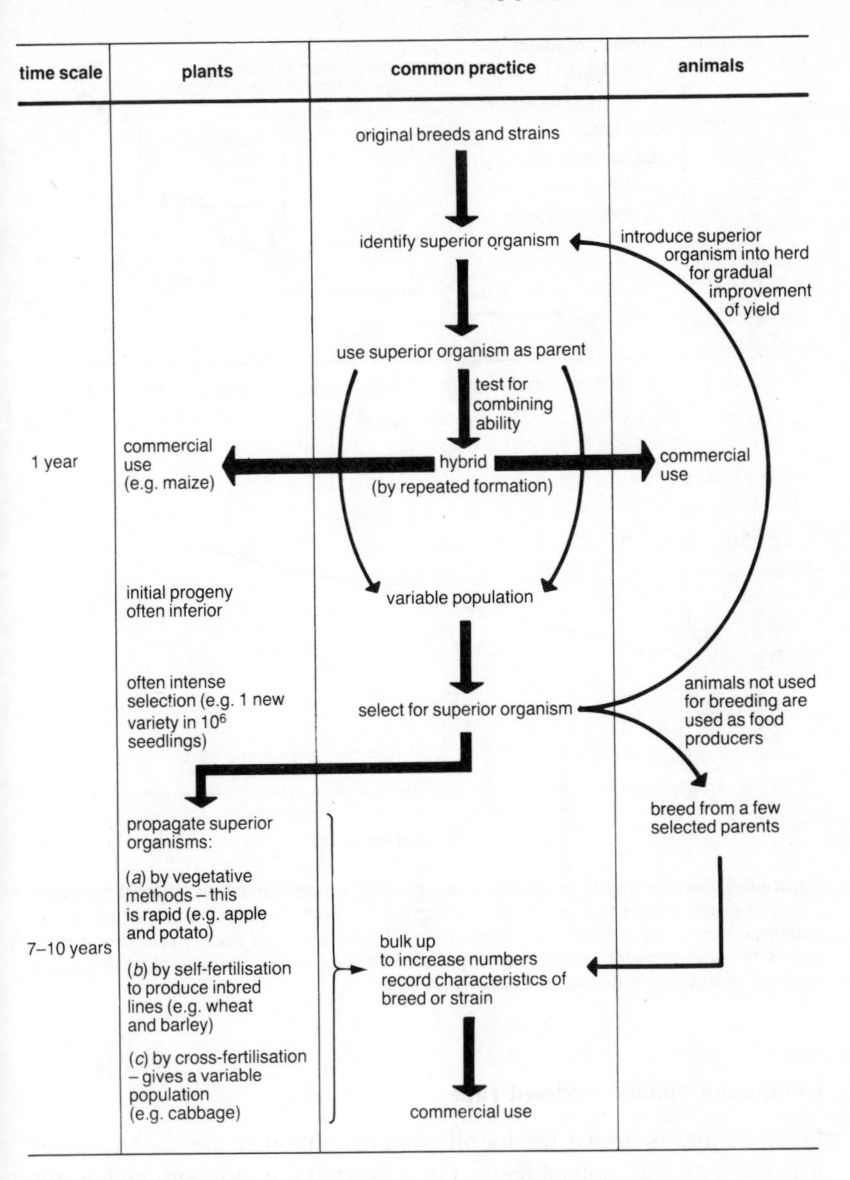

Figure 9.4 A summary of the ideas in Chapters 7 and 8: the flow of events associated with plant and animal breeding.

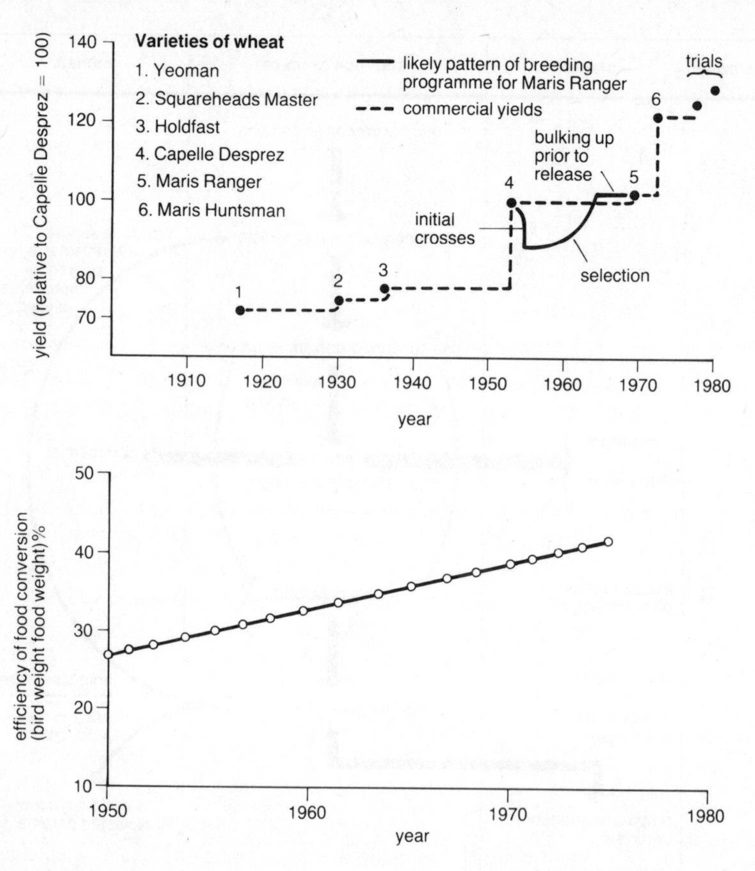

Figure 9.5 The patterns of progress for yield improvement in plant and animal breeding. The first graph shows the sequence of progress for plant breeding, using wheat as an example; this gives a step-wise improvement of one per cent each year. The second shows the sequence for animal breeding, using broiler meat as an example; this gives a gradual improvement of one per cent each year.

Combining quality – oilseed rape

Oilseed rape is grown for its oil content, although the seed residues provide nutritious animal feeds. For industrial use, oils with high levels of linolenic and erucic acids are favoured; edible oil for humans should ideally have low levels of erucic and eicosenoic acid and high levels of oleic and linoleic acid.

Canadian plant breeders were the first to produce a group of varieties, Canbra, which fulfilled the human nutritional requirements. Their breeding techniques were similar to those outlined in Chapter 8, but their assessment procedure was rather special. The cotyledon is the site of fat storage and therefore reflects the genotype of the embryo; one cotyledon (about 1 mg dry weight) can provide a sample large enough for the composition of the oil to be determined by gas–liquid chromatography, and yet the seed germinates and grows well after this surgery! Thus the phenotype of the plant could be determined in advance of the seed being planted, let alone grown to maturity.

Breeding experiments showed that the erucic acid content of the cotyledon was controlled by two genes at a single locus. Plants homozygous for one allele contained oils with only traces of erucic acid (less than 0.5 per cent) and very little eicosenoic acid (about one per cent), while the oils from plants homozygous for the other allele contained 50 per cent and ten per cent respectively of these oils. Heterozygotes contained intermediate levels of the acids, which is interpreted as independent, additive, gene action.

French scientists repeatedly backcrossed (page 121) Canbra strains to high-yielding varieties which lacked the Canbra oil composition. All the 'low-yield' genes in the Canadian variety were gradually eliminated, but not, of course, the alleles for favourable oil composition. The time taken for this backcrossing programme was reduced substantially by growing the plants in greenhouses. This made it possible to impose supplementary lighting, and hence allowed long-day rape to flower out of season; further, the chilling required by winter-sown varieties for vernalisation could be mimicked by storing the plants for several weeks in cold rooms.

This development of new strains of oilseed rape illustrates how changes may be made. First introduced into the UK on a large scale as a break crop (page 148) in intensive cereal production, it helped to reduce the incidence of disease in subsequent cereal crops and used existing farm implements for cultivation, sowing and harvest. Although the quality of the oil in the original varieties was not acceptable to the food industry, this was modified when the genetic control of oil composition was understood; an elegant breeding programme then led to increased cultivation of oilseed rape as a crop in its own right.

Increasing yield – short-strawed cereals

Cereal varieties with stiff short straws do not fall over when carrying heavy grains, following the liberal application of fertilisers. Also the division of biomass into grain rather than straw increases the yield as expressed by the harvest index.

Selection for short straw had taken place in the UK in the early part of this century. Straw height is determined by several genes, and progress was slow but steady. In the mid-1960s, however, a new range of semi-dwarf wheat varieties became available; this feature was incorporated into a variety released to the UK market in the early 1970s.

The source of this new genetic variation was Japan. As early as 1917, crosses were made between dwarf Japanese wheats and imported high-yielding American wheats; the variety 'Norin 10' became available in 1935, and strains were taken to the USA after World War II. In 1953 Dr Norman Borlaug, working for the Rockefeller Foundation, instituted breeding programmes to combine these semi-dwarf characteristics with high yield, disease resistance and the growth features required of a modern wheat. Techniques such as backcrossing were used. Strains for commercial use were available in 1963; in 1970 Dr Borlaug was awarded the Nobel Peace Prize for his contributions to world agriculture.

The reduction of straw length has been equally important in barley and rice breeding.

Improving quality – high-lysine maize

Plant foods contain small amounts of protein of generally poor quality. When they are used as food in simple-stomached animals such as pigs, poultry and human beings (page 31), they are thus of little value. Maize protein is no exception; ten per cent of the grain is protein but it contains insufficient lysine (an essential amino acid). A team of investigators at Purdue University sought to screen maize mutants for better protein quality.

In 1964 a high protein quality in a mutant called 'Opaque-2' was reported, kernels of which lacked the normal translucence. (The '2' refers to a previous cataloguing of opaque mutants.) But this was only the beginning; the Opaque-2 (o_2) gene had to be incorporated into high-yielding strains of maize, and its action understood.

The bulk of a maize seed is endosperm and this in effect is triploid: it derives from the fusion of a haploid male gamete with the two polar nuclei in the embryo sac of the ovule (page 79). At the locus controlling Opaque-2, all three alleles must be recessive ($o_2o_2o_2$) for high protein quality to be expressed. Individuals with genotypes $O_2O_2O_2$, $O_2O_2o_2$

Table 9.1 The inheritance cf opaque.

	pollen O_2	pollen o_2
polar nuclei $O_2 + O_2$	$O_2O_2O_2$	$O_2O_2O_2$
polar nuclei $o_2 + o_2$	$O_2\,o_2\,o_2$	$o_2\,o_2\,o_2$ (opaque)

and $O_2o_2o_2$ all have inferior protein quality. Gene action is dominant. The o_2 gene was incorporated into the local high-yielding strains by repeated backcrossing. (The genotype of normal, non-opaque, seeds could be established by self-fertilising the plants. There are two possible genotypes: the homozygous dominant individual (O_2O_2) yields only normal seeds, but the heterozygote (O_2o_2) yields some homozygous recessive individuals (o_2o_2) among its progeny, which exhibit the opaque trait. Opaque and normal seeds are in the ratio 1:3 (see Table 9.1)). Finally, the locus had to be made homozygous for the o_2 allele, which took five or six generations of further backcrossing and progeny testing.

High protein quality mutants of sorghum have been identified also, and the genes incorporated into high-yielding cultivars. Table 9.2 compares normal and high-lysine strains.

Table 9.2 Amino acid content of the grain content in high-lysine maize and sorghum (g/100 g protein).

	Lysine	Tryptophan
Normal maize	2.7	0.7
Opaque-2 maize	4.0	1.3
Normal sorghum	2.0	0.9
High-lysine sorghum	3.3	1.7

Breed replacement – milk production by cattle

Milk production by cows has been selected consciously and subconsciously so that several different breeds were created in different countries. The Dairy Shorthorn and Ayrshire were the most productive breeds in the UK until the middle of this century when the Dutch Friesian was shown to be superior; its use as a dairy breed then spread throughout the UK. Dutch breeders had been very successful at identifying and meeting the commercial factors which determine efficient milk production. The less productive breeds were replaced by more productive animals, a process called **breed replacement**. Many countries which adopted Friesians started their own programmes to improve the imported cattle. These programmes were based on **progeny testing**, in which the milking characteristics of the daughters determine the genetic value of the bull which sired them.

Programmes have been helped enormously by the introduction of **artificial insemination (AI)**, which enables a bull to sire many offspring – the semen is diluted before cows are inseminated (page 119), and semen is stored frozen, which extends the active life of the bull. The storage period enables us to obtain the milking characteristics of a sample of daughters before the semen is used. Preserved or newly obtained semen from proven bulls is released for large-scale insemination. The scheme is illustrated by the practice of the Milk Marketing Board of England and Wales (the MMB).

Bulls with the best daughters are mated to high-yielding and otherwise satisfactory cows. Each year, the MMB buys back about 150 bull calves from these matings; these are progeny tested. When they reach puberty, at about two years of age, semen is collected from each to inseminate enough cows to produce the 50 or so daughters needed in assessing the bull's potential. About 400 cows are inseminated, but only about 200 calvings are recorded; some animals do not become pregnant, and some farmers fail to monitor the outcome of these special inseminations. Half of the calves are female; these must be mated and their milk yield recorded. Some farmers leave the recording scheme, so only 50 or so progeny of the original 400 inseminations are actually tested.

The performance of these daughters is compared with that of their contemporaries in the herd; this 'contemporary comparison' is a test of the bull. 'Husbandry characteristics' of the daughters, which include general health and ease of milking, also help the MMB to decide which bulls to keep for AI service and which to cull.

The bulls are now seven years old and many have produced a sizeable quantity of semen, this being stored over the years. (In some countries, such as Norway, the bulls are not kept as live sources of semen; semen is collected from young bulls during the period of testing and all bulls are then culled. Much of the cost of keeping the bulls is

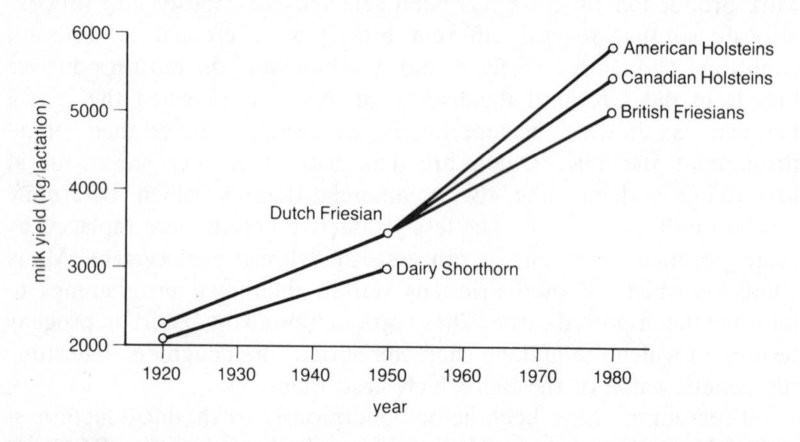

Figure 9.6 A representation of recent improvements in milk yield attributed to specific breed performance; half of the improved yield is due to genetic factors, the rest to improved environment.

saved, and the semen of bulls with low tests is simply thrown away.) The MMB estimate that the rate of improvement in milk yield is 0.5 per cent per year using their system. If they were to exclude husbandry characteristics, an estimated rate of genetic improvement of yield nearer two per cent could be achieved.

As regards the future, the North American countries (USA and Canada) have been very successful in improving their national herds. The improvement between 1960 and 1975 in the genetic merit of their sire studs amounted to 600 kg of milk per lactation, the original yield being 4200 kg of milk. The rate of genetic improvement (around 40 kg per year) over this period is twice that shown in the UK. (Because we have also improved the environment in which the animals are reared, the actual improvement in yield in both the UK and North America is about double this, as shown in Figure 9.6.)

Breed replacement is again taking place. Genes of the British Friesian herd are gradually being replaced by those of American Holsteins, the process being hastened by AI. The sequence of events in the improvement of milk yield is summarised in Figure 9.6.

Altering breeds – pig carcass quality

As with dairy animals, the British pig of today is very different from that of 30 years ago. Present-day animals have more muscle and less fat. Lean meat is what the consumer wants, so the pigs have higher carcass quality. The change in carcass quality took place by a different process from that of dairy animal improvement, because the pig breeder did not have another breed available; all breeds of pigs were too fat. The new demand for lean pigs had therefore to be met by genetic selection to remove fat.

The depth of fat on a live pig can be measured using a special ultrasonic scan. Thus the pig breeder can identify lean pigs and use these as parents rather than their close relatives (as would normally be true of a carcass characteristic). The depth of back fat on pigs in Britain has declined from 25 mm in 1950 to under 15 mm today – such is the power of animal improvement programmes.

Figure 9.7 illustrates the change in back fat obtained by Dr Nils Standal and his colleagues in Norway who selected for back fat alone over a ten-year period. Starting at around 20 mm they could increase it to 35 mm or decrease it to 15 mm by genetic manipulation alone. You can decide which chop you would prefer!

The change in British pigs has been less rapid because other features were included in the selection index (page 128); back fat was just one. The progress of selection was slower, but because other characteristics improved also, the overall merit was greater.

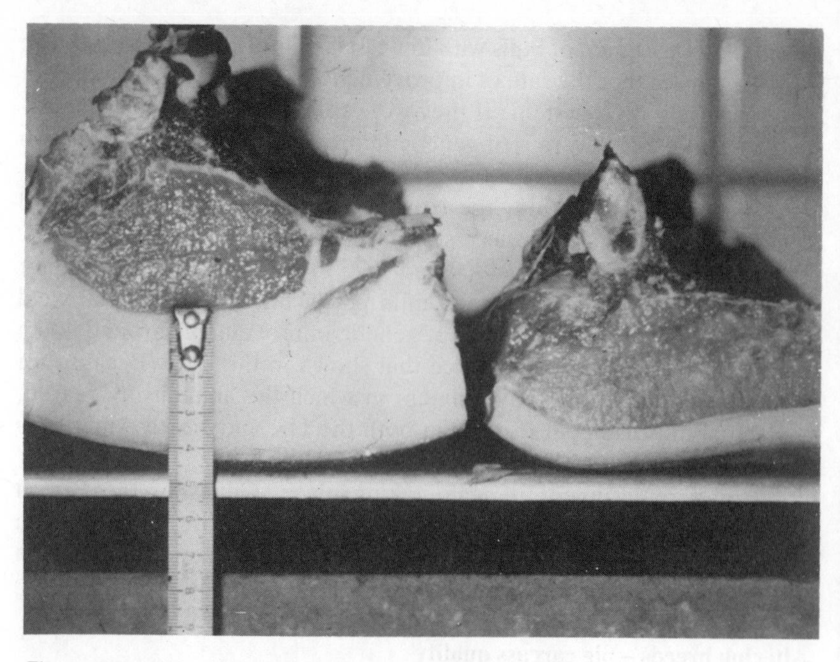

Figure 9.7 Joints of meat from fat and lean pigs.
(Courtesy of Dr Nils Standal.)

One trait important to the economics of producing pigs is the efficiency with which food is converted into body weight. An improvement of four per cent in food conversion will save 7 kg of food during growth of the pig from 20 kg to 90 kg live weight. Progress in decreasing the back fat and in efficient conversion of food into meat means that only 2.5 kg food, instead of up to 3.5 kg food is now needed to produce 1 kg of live weight; a change engineered in the last 25 years.

The Meat and Livestock Commission of the United Kingdom has been responsible for many of the programmes of pig improvement. It has provided testing facilities to monitor food conversion (Figure 9.8) and carcass quality, and created selection indices. During 1970–80, the cost of production in Britain fell by 70–100 pence per year per pig. Since some ten million pigs are produced each year, the overall saving was about £100m per year, with an annual investment of only £2m in research.

British pig breeders have been so successful that superior pigs have been exported to many parts of the world. The breeders' policy is to select within the two major breeds, Large White and Landrace, and then mate them together to exploit the heterosis (page 99) which such a union shows. A hybrid dam is usually mated to a pure-bred sire and the offspring grown up for meat.

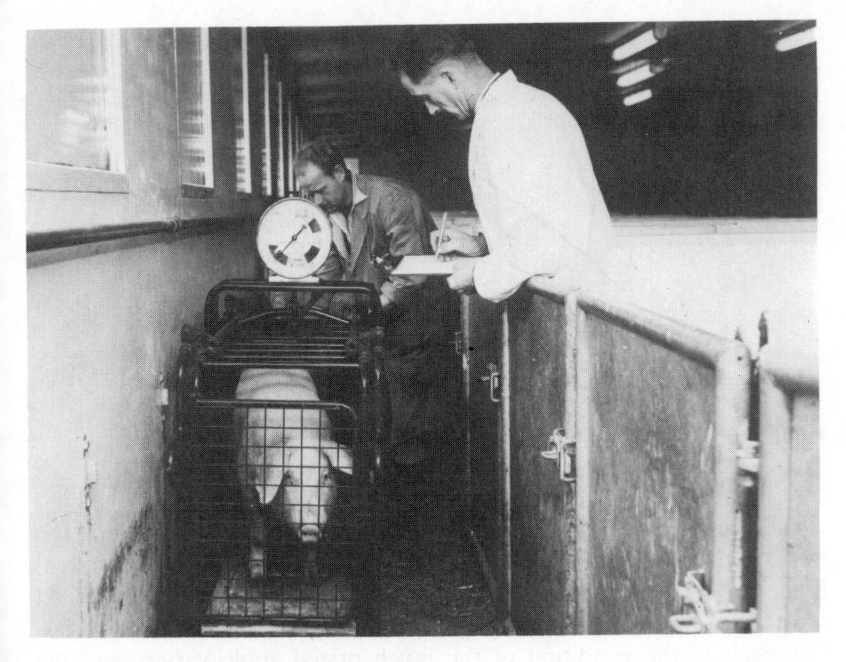

Figure 9.8 Pig performance testing facilities.
(Courtesy of the Meat and Livestock Commission.)

Success and future problems – poultry breeding

Thirty years ago, poultry meat was considered a luxury, but planned genetic improvement has had a greater impact on poultry production and acceptability to consumers than on any other livestock. Poultry breeding has been left to private enterprise, so details of progress, programmes and performance are not readily available.

Producers of poultry for meat (broilers) or eggs (layers) have been concerned mainly with the efficiency of converting food into these products. Broiler birds have been selected for rapid growth, to reduce food and housing costs; layers have been selected for small body weight, to reduce the amount of food needed for maintenance.

The successes of poultry breeding also illustrate some of the problems of genetic improvement. It is argued that genetic selection is now having little effect on egg production (Figure 9.9). Has genetic variation been exhausted? This is difficult to answer. Efficiency of production is affected by costs as well as output. Current bird strains may not produce more eggs, but they make fewer demands on the environment than did their predecessors – they eat less food and are less prone to disease. So, in fact, genetic progress does continue to improve the efficiency of egg production.

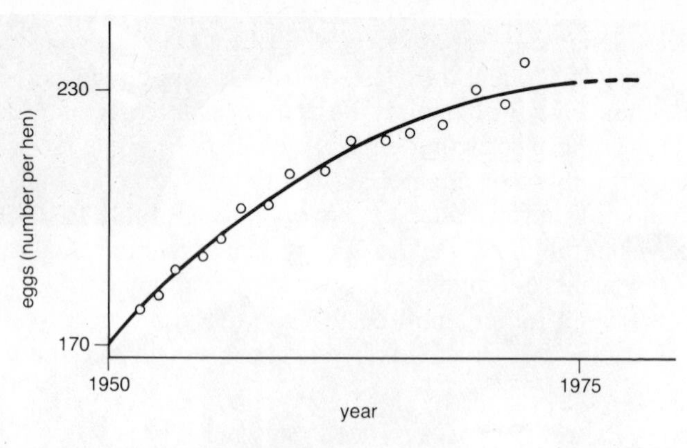

Figure 9.9 Trends in egg production.

Selection for weight gain in broiler poultry has made the meat too fat, so selection must now be made for leanness and food conversion. Some turkey strains are so large that they cannot easily mate, and AI must now be used to propagate the strain. The animal is fast approaching the condition of the much prized apple variety and the biologist must ensure that it is propagated; but grafting will not solve this problem!

The economic return from improvement

In essence, all improvements to the yield of food-producing plants and animals are made in order that there be profit, usually financial profit. Even aesthetic improvements can be measured in financial terms; likewise, old and apparently unproductive strains or breeds (such as Longhorn cattle) are maintained to serve a market. And some species are maintained to provide a 'gene pool' for existing or future breeding programmes.

In strictly financial terms, we can define the profit of food making as

$$\text{profit} = \text{value of output} - \text{cost of production},$$

Table 9.3 Costs of food making.

Fixed costs	**Variable costs for plants**	**Variable costs for animals**
rent for land	seed	animal replacements
buildings	fertilizer	food
labour	fuel	fuel
machinery	crop protection	veterinary charges

and we can define

cost of production = fixed costs + variable costs

This book has dwelt mainly on the *variable* costs of production because it is in these costs that improvement can be made most readily. Variable costs apply to the materials we use to make food; they depend on the biological performance of our organisms (Table 9.3).

Changes in the *fixed* costs of food production entail changes in the way we grow plants or animals. Thus we might rear animals out of doors all the year round rather than mostly indoors; or we might substitute the use of Paraquat (page 58) as a 'chemical plough' for the use of a conventional plough.

The law of diminishing returns

This law was discussed in Chapter 5, but we need to review it in economic terms. Figure 5.1 showed that for increasing quantities of a component of the feeding environment, the extra amount of food produced decreases. This component, if we supply it, is likely to be of variable cost, and its cost may be imposed upon the graph. The maximum *economic* yield is often below the maximum *possible* yield, and farmers naturally favour the maximum economic yield.

New varieties and breeds

For a new variety to be introduced, it must offer the producer higher profit. Figure 9.10 summarises some of the ways that this can be achieved in theory.

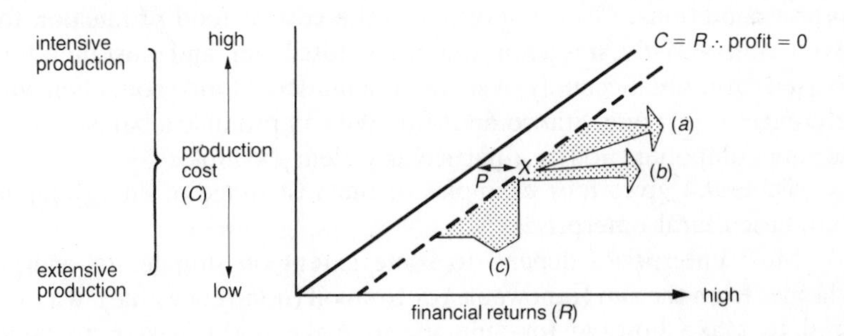

Figure 9.10 Ways of increasing profit from an agricultural enterprise. There are three types of improvement: (*a*) higher costs, but even higher returns (e.g. straw length, see page 133); (*b*) same cost, but higher returns (e.g. poultry energy conversion, see page 139); (*c*) lower costs and higher returns (e.g. disease resistance, see page 147). (After Simmonds.)

How does this work in practice? For the new *plant* strain this is relatively simple. New superior strains appear infrequently (page 132), but the superiority is usually obvious, so producers are willing to grow it. The plant breeder releases large quantities of seed to the seed merchant who produces more seed and so makes it possible for many farmers to introduce the strain.

By contrast, the superiority of *animals* is difficult to demonstrate – farmers are rarely convinced that a particular sheep eats two per cent less grass. Superior animals are produced in isolation and improvement is gradual (page 132). The breeder may have to convince two groups of farmers that an animal is superior: those who use the animal to improve their own flocks, and those who buy animals bred from the new programme. Yet the gradual improvement of animal breeds, as well as plant breeds, is important to the national economy – just as a two per cent saving of food in intensive poultry production for an individual farmer may decide between success and bankruptcy.

Oil – a variable cost

In early 1974 it became clear that the cost of oil would substantially affect any food-producing enterprise, partly by increased cost of fuel, and partly by increased costs of raw materials needed in the manufacture of equipment and of chemicals such as fertilisers, veterinary drugs and crop sprays. World inflation started its upward spiral at about the same time.

When oil prices increase, so do food prices. Economists calculated how much energy is used in producing certain types of food under certain conditions. Oil is important in the cost of food production for two reasons. Firstly, it is a non-renewable fossil fuel, and must therefore be used carefully. Secondly, it serves as a model of food production with reference to just one vital component. We can produce a balance sheet for any component; oil is a particularly clear example.

Table 9.4 gives four examples of the vast range of energy inputs into agricultural enterprises.

Most enterprises depend to some extent on supplies of energy, whether from the sun (renewable) or from oil (non-renewable), whether used to make housing for animals, to make herbicides or to 'feed' humans, animals and tractors. One eminent scientist has suggested that we owe the higher yields of present-day agriculture largely to the fact that we do not feed horses! The oil to power tractors has released agricultural production to feed humans.

Intensive production of plants and animals does not employ much human labour, but is expensive in other energy resources; extensive

Table 9.4 The energy balance sheet for intensively and open range (extensively) produced plant (maize) **a** and animal (beef) **b** foods. All units in kJ ha^{-1} y^{-1} unless otherwise stated.
(Based on Pimentel and Pimentel, 1979.)

a

	Total energy input, including labour, seed, machinery and oil	**Energy input as fossil fuel**	**Total energy output**	**Actual yield (kg ha^{-1} y^{-1})**
Intensive	27×10^6	25×10^6	80×10^6	5400
Extensive	3.4×10^6	0	14×10^6	940

Total light energy 6×10^9

b

	Energy input as food	**Energy input as fossil fuel**	**Energy output**	**Actual yield (kg ha^{-1} y^{-1})**
Indoors	10×10^7	7×10^7	4×10^5	300
Open range	6×10^7	0.5×10^6	1.6×10^4	12

Total light energy 6×10^9

production relies on human labour with little oil input. Plants are efficient providers of food; the energy yield exceeds the non-renewable energy input. For animals, food energy input must exceed energy output; this aspect of food making was summarised by the second law of thermodynamics (page 6). It states that energy is inevitably lost as heat when food passes from one member of a food chain to the next.

The government and marketing

The world is not economically perfect; governments dictate the supply and demand figures to ensure a politically and economically satisfactory balance within the food industry and between the food industry and others.

Regular reviews of pricing and trade agreements seek to provide a stable economic climate for producer and consumer. Political decisions sometimes cut right through biological logic. Two examples will illustrate this. When Rhodesia illegally declared independence in 1965, the United Nations declared trade embargoes. Rhodesian tobacco, for example, could not be sold, and Rhodesian farmers had little by way of legal income. Likewise, the Soviet people and their livestock were deprived of grain following the invasion of Afghanistan, while North America had vast surpluses and producers suffered reduced income.

Since the economic and social disruption of World War II, various producer groups have united against over-production, while consumer groups have done likewise against under-production. Several cartels (associations of independent producers) operate internationally to control prices and distribution of commodities such as coffee. Merchants establish 'future markets' in which consumers can buy a commodity for delivery several months hence. These help to provide price stability, but allow fears of over-production or of shortage to be whipped up into almost panic speculation.

Under certain circumstances, producers and consumers join forces to provide stable market conditions. Perhaps the greatest such association was the Commonwealth; when rid of Imperial exploitation and united by mutual difficulties, this served as a marketing union. The sparsely populated agricultural economies found ready markets for their food in the highly populated countries, and the industrial areas found markets for their products. This state of affairs did not remain stable; and primary producing countries began to satisfy their own needs for industrial goods.

In the wake of World War II, the European countries sought to create a marketing union. Centred on steel and coal at first, it now encompasses all trade. The European Economic Community (EEC) as we know it was created in March 1957 by the Treaty of Rome, and came into being in January 1958; four other countries have joined since then.

The agricultural aspects of the EEC alone illustrate the nature of this type of marketing structure. In essence, politicians set the prices of certain goods. Prices are reviewed annually and at all times goods should flow freely from one country to another. However, the price farmers receive for their goods is calculated in 'units of account' (ua) which can be manipulated. In 1974, £1 was equivalent to 2.16 ua; in November 1975, it became equivalent to 1.76 ua; and in January 1981 to 1.34 ua. This exchange rate is determined by the politicians.

In theory, the farmer is assured a basic price for his goods: the **intervention price**. If the price of a commodity on the international market falls below this price, then the funds of the community are used to buy the commodity from the farmer, provided that its quality is high and that fairly large quantities can be dealt with. The price structure caters for any commodity which the EEC itself over-produces.

In practice, problems arise because agricultural prices are linked to units of account ('green money'), while the farmer's expenditure (wages, fertilisers, machinery and so on) is all in real money. Differences in the actual *value* of the green currency mean that producers in one country can sometimes underprice producers in another country, because of the free movement of goods across national frontiers. An attempt to simplify the agricultural pricing of the EEC has actually resulted in complex problems, such as the various 'lakes' and 'mountains' of foodstuffs now associated with the EEC's Common Agricultural Policy (CAP).

So we come full circle in this chapter. The examples of successful breeding programmes to improve yield may have left the impression that the biologist is firmly in control of plant and animal improvement programmes. Doubt must have arisen at the mention of economics; and with the introduction of politicians, all hope of biologists controlling events must have vanished. As we shall see in Chapter 12, onerous responsibilities are placed upon biologists; demands will continue to be placed on their willingness to respond rapidly to changed political and social patterns.

Summary

The parts of an organism which we use constitute its yield. Improvement tends to follow this pattern:

a identify the yield in the context of biological and economic objectives;

b maximise the yield by a suitable breeding and selection programme;

c market the product in a local or international economy.

10 Disease

Disease often plays an important part in the choice of methods of food production. Micro-organisms have vast reproduction rates (Chapter 6), and are thus ideally suited to the opportunist role of a disease organism. Reduced yields and the extra cost of disease control (Chapter 9) may erode the financial return from food production. In this chapter, we turn our attention to disease: to the biology of its spread and to the ways in which we can control this.

Several approaches may be adopted. One is to identify existing genetic resistance in some individuals and to introduce this into a new strain (Chapter 8). Natural enemies restrain the spread of a disease; these can be exploited. Chemicals may be particularly harmful to pathogens, but the 'broad-spectrum' kind may also contaminate *our* feeding environment, so they must be used sparingly and with caution. Chemicals which are toxic to specific pests and which are broken down rapidly after application, may be essential to pest control.

Our ultimate aim is to develop a pest-control programme with an integrated biological approach, where all possible natural and artificial controls are used with the minimum disturbance to other organisms in our feeding environment.

What do we mean by 'disease'?

We may divide diseases into those caused by agents (such as a virus, bacterium, fungus, protozoan, roundworm, platyhelminth or arthropod) and those caused by physiological factors. This chapter deals largely with parasitic agents, large and small, plant and animal, external and internal. ('Nutrition' or 'production' physiological diseases were discussed in Chapter 5.)

Types of host and types of disease

There are two extremes of host life style. On the one hand there is the 'opportunist'; growth is rapid, and a whole life cycle, from egg to adult, may take place in only half of a year. Intensively reared poultry and tropical crop rice are examples. On the other hand, there is the 'plodder'. Large farm animals and trees grow to full sexual maturity only after a long period of time; an apple orchard may take a decade, and cattle two or three years.

Associated with these extremes of host life style are characteristic agents of disease. A game of hide-and-seek exists between the short-lived (ephemeral) host and *its* pathogen. With these aggressive and rapidly evolving diseases, our action must be decisive; chemicals are applied and resistant strains are bred. At the other extreme, food losses from the longer-lived organism may often be avoided by ecological methods which moderate the pathogen's growth, so that only occasionally is it necessary to use chemicals.

Growth rate of disease agents

The increase in pathogen numbers after infection follows the pattern discussed in Chapter 6, and thus the spread of a disease generally follows the S-shaped curve of Figure 6.3. The rate of increase declines as the food supply becomes restricted by the severe incapacity of the host. Even before this stage of infection is reached, poor growth of the host or damage to it which makes it unsaleable has reduced *our* food supply.

The speed of infection varies. Pathogens in 'opportunist' hosts must in turn be 'opportunist', and the increase in their numbers must be rapid. Potato can be infected by potato blight (*Phytophthora infestans*), and the yield significantly reduced, within a very few days; whereas the increase in numbers of the liver fluke (*Fasciola hepatica*), in a flock of sheep, follows the plodding course. Long-lived hosts may be infected by both 'plodder' and 'opportunist' pathogens.

Methods of infection can be very important. The spores of potato blight are dispersed by the wind and require a suitable host and a suitable environment – particularly with regard to humidity – to germinate, grow and spread further. Continuance of the liver fluke requires its larvae to be deposited on the blades of grass and hence ingested by grazing sheep; the fluke is dispersed when infected sheep contaminate the grass which disease-free sheep then eat, rather than by mobility of the infective stage.

We can couch the speed of these growth rates in mathematical terms, and so appreciate the factors involved. Because the increase in numbers is exponential, the number of pathogens n at time t will be

$$n = n_0 e^{rt},$$

where n_0 = the initial number (the size of the initial infection),
e = a constant, the base of natural logarithm 2.72,
r = the relative growth rate (the 'rate of interest'),
t = the time of sampling (how long the infection has been allowed to continue unchecked).

Prevention is better than cure. There are two principal approaches to reducing the losses from disease; both are based upon our biological knowledge of the disease agent and its environment. One way is to *avoid* disease, the other is to *control* it.

Prevention – hygiene and resistance

Of the various ways we can prevent the spread of a disease, hygiene (whereby the initial number of disease agents is reduced) and resistance (whereby the host itself reduces the speed of invasion) are perhaps the most effective.

Specificity of disease

The most successful way of preventing the spread of a disease is specific resistance. We take this for granted, yet how much greater our food losses would be if every food producer could be infected by every disease agent. We never even consider foot-and-mouth as a potential disease of lettuce, or potato blight as a disease of chickens. Most pathogens have a *specific host range*; this may be limited to just one species or may extend to several. *Botrytis cynerea*, for example, produces grey mould on grapes, lettuce, tobacco and strawberry.

Species which are not susceptible to local diseases may be specially cultivated. Where there is a high incidence of one disease agent in a given area, we can grow a different food organism which is not susceptible to that agent. This is a form of hygiene; we do not allow the host to make contact with the disease. Thus pastures may be 'rested', allowing animal parasites to die naturally. Similarly, the use of a so-called 'break crop' allows the soil to be used, but by a species which is not susceptible. The increased plantings of maize in traditional wheat and barley areas in the south of England are an example; maize is not susceptible to nematode diseases or to fungal diseases such as 'take all' and 'eye spot', and so farmers can plant maize rather than leave the field fallow for a year.

In terms of our formula, $n = n_0 e^{rt}$ (page 148), hygiene makes $n_0 = 0$, and resistance makes $r = 0$ also.

Strain resistance

A pathogen's host range can be restricted not only to individuals of one species, but also to only some of the individuals within a species. Certain strains or cultivars of a species may resist particular disease, while others remain susceptible. These differences can be attributed to two types of genetic control.

In the first type, a single **major gene** in the host (an *R* gene) matches a particular gene in the parasite (resistance is said to be pathotype-specific). The marked differences between strains in resistance to a particular disease are often called **vertical resistance** (**VR**). This system requires some form of recognition, perhaps of a protein produced by the particular gene in the parasite. When resistance is shown by the host, growth of the parasite is checked; the parasite is fiercely selected against. Any individual parasite in which this particular gene mutates to give an unrecognisable protein has a selective advantage; it will live and reproduce at the expense of others which do not have this mutation. Gene frequencies may change rapidly and a population explosion may result – hence the so-called 'boom-and-bust' cycles which sometimes occur when the plant breeder introduces a new *R* gene into a commercial variety. Such a cycle occurred when the fungus 'yellow rust' (*Puccinia striformis*) suddenly altered its genotype and infected the new and promisingly resistant wheat varieties 'Rothwell Perdix' and 'Joss Cambier'. These strains are no longer grown.

There are some reports in animals of a link between disease resistance and the activity of a single gene. Breed differences exist between sheep; Masai sheep are more resistant to *Haemonchus* worm than are Merinos. And individuals within these breeds which have the B haemoglobin type (Hb^B) are more susceptible than those with the A type. Blood protein types in sheep, like those in man, are controlled by genes at one locus; the *A* allele is responsible for the A blood group as the *B* allele is for the B type. Thus the specific difference between individuals stems from one gene; also this gene may affect the general robustness of sheep (lambs with the A type survive better in the Scottish hills than do those of the B type). As yet, we cannot explain these effects.

There is a similar example in man: the 'Duffy' blood group. This system is controlled by two out of three possible alleles (Fy^a, Fy^b and

Fy^o) at the locus (as with the ABO system). Individuals with either the *Fy*^{a} or *Fy*^{b} allele are susceptible to malaria (*Plasmodium falciparum*); those without either (*Fy*^{o} *Fy*^{o}) are comparitively resistant. Not surprisingly, high proportions of some native African tribes have neither *Fy*^{a} or *Fy*^{b} alleles, whereas Caucasian (white) populations have one or the other.

The second type of genetic control depends on many genes and this gives **field** or **horizontal resistance (HR)**. Differences in resistance may be small between strains. The resistance is analogous to a combination lock; to open it, you must get perhaps five numbers correct; similarly, the parasite must alter *several* of its genes to be able to break down the host's resistance. This polygenic system is efficient at maintaining host resistance, but is very difficult to manipulate in a breeding programme.

Despite these difficulties, plant breeders have turned their attention towards incorporating field resistance into crops such as the potato, to combat potato blight and so avoid the uncertainty of 'boom-and-bust' cycles.

Breed differences do exist in animals, but differences have been recorded only recently and are not yet fully understood. Zebu cattle (*Bos indicus*), which thrive in the tropics, are more resistant to the cattle ticks which so readily infect European-type cattle (*Bos taurus*) in the tropics.

At the species level, sheep and goats are less susceptible to trypanosomiasis than are cattle. As the environment becomes increasingly favourable for the fly vector, it is the cattle which are lost first from a mixed population of cattle, goats and sheep. As differences in sensitivity are understood, breeders may try to incorporate the advantageous features into higher-yielding animal breeds.

In seeking to improve resistance of the host to a parasite, any pattern of control must be based on an understanding not of the genotype of the host alone but of the parasite also. Breeding programmes in the host must be directed in part towards anticipating the genetic capability of the parasite. (The genetic systems of parasites are beyond the scope of this book.)

Returning to our formula, $n = n_0 e^{rt}$, we can see that strain resistance can make $r = 0$.

The nature of resistance

Plants and animals differ greatly in the way that they deal with a disease agent. All vertebrate *animals* grown for food have the same basic antibody system, a system which recognises the chemicals in the parasite and produces complementary chemicals (**antibodies**). Host and parasite chemicals when locked together inactivate the parasite, which is then removed by other means. The relationship between the speed of induction of this antibody reaction to any antigen and the likely resistance to infection by a particular antigen is of great interest. Unfortunately, however, the results of experiments to date are inconsistent. A large response by the host *may* indicate susceptibility, a need to react positively, or it *may* indicate resistance; antibodies so formed rapidly inactivate any antigen!

In contrast with animals, *plants* have no such antibody reaction, but are by no means defenceless. Disease is checked by inhibiting either infection or subsequent growth of the parasite. Infection tends to be controlled by major genes in the host, which shows vertical resistance; while growth rate is normally governed by a polygene system, and is horizontal resistance.

Inhibition of infection often provokes a hypersensitive reaction in the host; this is especially true of fungal diseases. The disease agent penetrates the physical defences of the host, but a rapid local reaction produces a patch of dead host cells which surround and isolate the disease agent; these lead to tiny necrotic (dead) patches. Sometimes phytoalexins (chemicals which belong generally to the phenol and terpenoid groups of compounds) are involved; these are fungicidal or fungistatic in action. They are produced in response to natural infection of the host by pathogenic organisms, as we might expect, but they are produced also when the plant is inoculated with a non-pathogenic organism which would never invade the host naturally, and when certain inorganic ions such as copper are applied (see page 154). About 30 phytoalexins have been isolated, the first being pisitin, isolated from peas in the early 1960s. They are now thought to be produced as a general response to damage.

Chemical control

Until quite recently, food producers believed that chemicals were essential in controlling disease so as to reduce food losses. Chemical companies therefore produced chemicals with a broad spectrum of activity. Initially there was little regard for their modes of action or concern for their acute (short-term) toxicity to the user or for their persistence in the ecosystem. We now realise that chemicals can show chronic (long-term) toxicity to the consumer, so persistence in the environment must be short. We are particularly vulnerable to this form of poisoning because we may be the secondary (or later) consumer. The chemical is concentrated as food progresses along the food chain.

A second phase in pesticide development is now well under way; tailor-made, low-toxicity chemicals with specific short-term activities will become increasingly widespread.

Resistance to chemical treatment

When a toxic chemical is applied to any organism, the chemical kills the disease agent or greatly reduces its growth but the environment is contaminated. However, any disease agents which survive because they are genetically better suited to resist that chemical pass on this resistance to their offspring; there is a sudden change in gene frequency (page 101). (We saw this when disease organisms overcame the genetic resistance shown by the host.) Thus the resistance spreads, creating a new 'super breed' – as happened when rats became resistant to Warfarin (a blood anti-coagulant) and when some insects became immune to the insecticide DDT. The effectiveness of the chemical is reduced, and its use should then be discontinued.

Insecticides

Many chemicals act as **insecticides** on plants or animals. DDT, perhaps the best-known insecticide, has been used equally effectively against insect pests on plants, domestic animals, wild animals and even humans. It was first used in Naples in 1943, when it prevented an outbreak of typhus among humans by killing the louse which carried the disease. Since then an estimated 1.5 million tonnes have been applied over the earth. Lessons learned from DDT serve as a guide for the development of new insecticides; the major problem stems from its persistence.

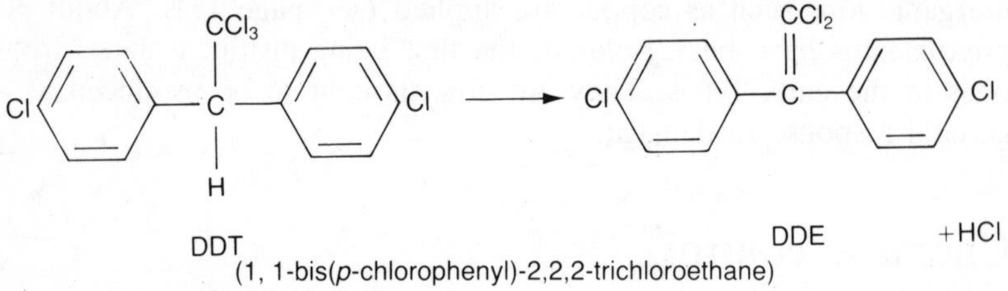

DDT (1, 1-bis(*p*-chlorophenyl)-2,2,2-trichloroethane) → DDE + HCl

Figure 10.1 The molecular structures of DDT and of the first of its breakdown products, DDE; both have insecticidal properties.

DDT (Figure 10.1) introduced into our ecosystem is concentrated naturally. Organisms in the food chain become contaminated, and because DDT is fat-soluble, animals store it in their fat reserves rather than excreting it. DDT also has a very slow rate of breakdown. Thus it has persisted and it has been possible for resistance to evolve. (Unfortunately even when DDT does break down, another persistent chemical DDE is formed; this also has insecticidal properties.)

How do newer insecticides compare with the almost indestructible DDT? One insecticide, pyrethrum (a compound extracted from the flowers of a plant belonging to the chrysanthemum family), was first extracted and used in 1828 and is still used in domestic fly sprays because of its low toxicity to us.

The so-called **chlorinated hydrocarbons** (the group which includes DDT, lindane, aldrin and dieldrin) persist in a biologically active form for up to ten years. These have been replaced by the non-persistent **organophosphorus compounds**, and Figure 10.2 shows that when one such compound (parathion) breaks down, simple non-toxic products are formed.

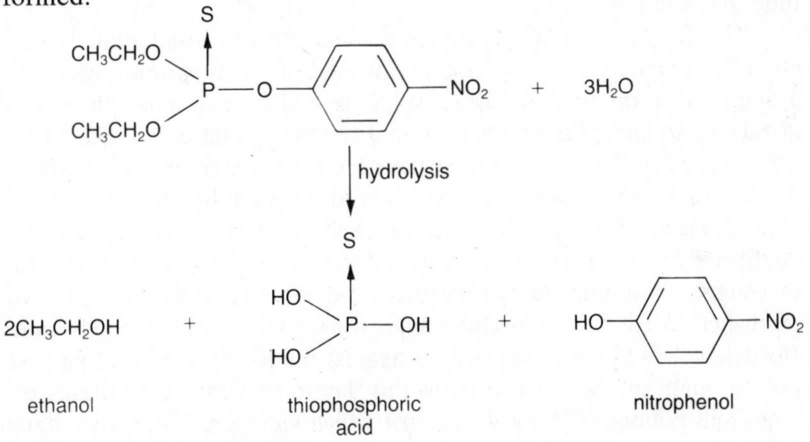

S
CH_3CH_2O
P—O—
NO_2 + $3H_2O$
CH_3CH_2O
hydrolysis
S
HO
$2CH_3CH_2OH$ + P— OH + HO— —NO_2
HO
ethanol
thiophosphoric acid
nitrophenol

Figure 10.2 The molecular structures of parathion, an organophosphorus insecticide, and of the compound formed when it is hydrolysed.

The organophosphorus compounds are also **systemic** (that is, they can enter a plant at one site and be effective wherever needed). They protect new growth from insect damage and the chemical is protected by the plant from the weather.

Despite the variety of organophosphorus compounds, they seem to act on the insect in basically the same way. They interfere with the passage of nerve impulses from one nerve cell to another. Impulses may be transmitted across the synapse by acetylcholine which is then destroyed by an enzyme, acetylcholinesterase (ACHE). The compounds occupy the active site of the ACHE molecule so that acetylcholine is not broken down. The resulting accumulation of acetylcholine causes a stream of impulses to be passed down the nerve cells or axons – which explains the unco-ordinated movements which accompany the insect's death.

Fungicides

The origins of fungicides, like those of selective weedkillers, were empirical; early selective fungicides were inorganic substances such as sulphur and copper salts. Both of these were applied first to the vine.

Sulphur was first used in 1843, against powdery mildew in an English glasshouse. It is still used; finely ground sulphur is applied as a dust three or four times a year (about 70 kg/hectare). For dry applications, an inert 'filler' such as gypsum is used to dilute the sulphur so that leaves are not scorched on hot days. For wet applications, sulphur is applied as a spray of sulphur-lime, which was originally made by mixing lime and sulphur.

The original use of copper as a fungicide was accidental. It was common practice in the Médoc region of France to sprinkle roadside grapes with a mixture of copper sulphate and lime to make them less attractive to the pilferer. A certain Professor Millardet noticed that grapes growing by the road held their leaves until the end of October, while other vines had by then been defoliated by downy mildew (*Plasmopara viticola*). This was in 1882; in the following year he confirmed his observations. In collaboration with a Professor Gayon, he investigated the fungicidal properties of the mixture; their findings were published in 1885. The mixture, known as Bordeaux mixture (Bouille Bordelaise), was used successfully against mildew in 1887 and against potato blight in 1888; it is used to this day to prevent fungal attack on vines and potatoes. The walls against which vines grow in France have now a characteristic blue copper colour – an indication of how much has been applied.

Modern fungicides can be either contact or systemic. Their chemistry and modes of action are complex.

Ecological control

We have noted that chemical control has drawbacks, such as contamination of the ecosystem; it is desirable to be able to control pests without resorting to chemical methods. By manipulating the biological environment of the organism, and interfering with its chance of survival or reproduction, we can out-manoeuvre the pest.

Control by predation

The pests of our food-producing organisms are rarely at the end of a food chain; some organisms can use them as food. If we know the biology of the pest and of its predators and parasites, we can encourage the predators. But in the same way that we must be sure of the specificity of chemicals, we must be sure that the predator will not attack other organisms – organisms which stabilise our ecosystem or form part

Table 10.1 Examples of ecologically based pest control.

Damage caused	**Damaging organism**	**Biological control organism**
greenhouse crops in UK	red spider mite (*Tetranychus urticae*)	mite (*Phytoseiulus persimilis*)
	whitefly (*Trialeurodes vaporariorum*)	wasp (*Encarsia formosa*)
arable and grass crops	rabbit	myxomatosis
sugar beet and rose	aphid (*Myzus persicae* and *Aphis fabae*)	adult and larval stages of 5- and 7-spot ladybirds and larvae of hoverflies
agricultural grazing	prickly pear (*Opuntia stricta*)	Argentinian moth borer (*Cactoblastis cactorum*)
mammals	mosquito (vector, spreading malaria)	fish (e.g. *Gambusia affinis*) eats mosquito larvae
spruce forests	European spruce sawfly (*Neodiprion sertifer*)	nuclear polyhedrosis virus (PNV)

of our food. Further, the predator must be able to reproduce and grow rapidly under those conditions which favour the pest. The potential of this so-called **biological control** is enormous, largely because it is self-perpetuating. Many examples are known; Table 10.1 illustrates the diversity of control which we can exercise.

Control by other methods

The term 'biological control' is used to mean control by predation, but there are other methods of control which exploit the biology of the pest. Organisms are vulnerable to regulation when they reproduce and when they feed, so biological methods seek to reduce the reproductive and growth potential of pests.

Male sterilisation is one very successful way of reducing insect pest numbers. Sterile males are introduced into the population and allowed to mate with normal females. The method is successful because insects mate only once; the female then stores a package of sperm. If these sperm are infertile, the numbers of offspring will fall. A requirement of this method is that the infertile male must behave like a fertile one, mate as normal and produce motile sperm, but sperm which is defective and does not allow normal embryo development. Clearly this method is limited to organisms which mate only once in their reproductive lives –

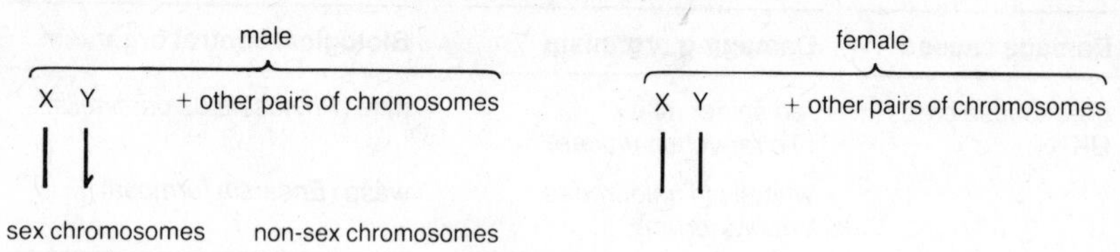

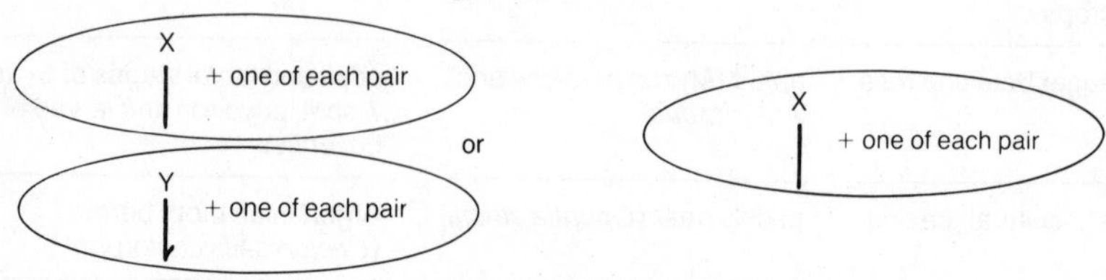

automatic sexing system

1 Find a locus for insecticide resistance. At the *R/r* locus:

RR and *Rr* give resistance, *rr* gives susceptibility.

2 This locus must be on a pair of non-sex chromosomes (say, the A chromosomes):

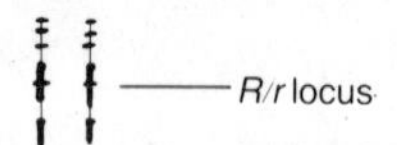

3 Effect a translocation between the Y chromosome and an A chromosome:

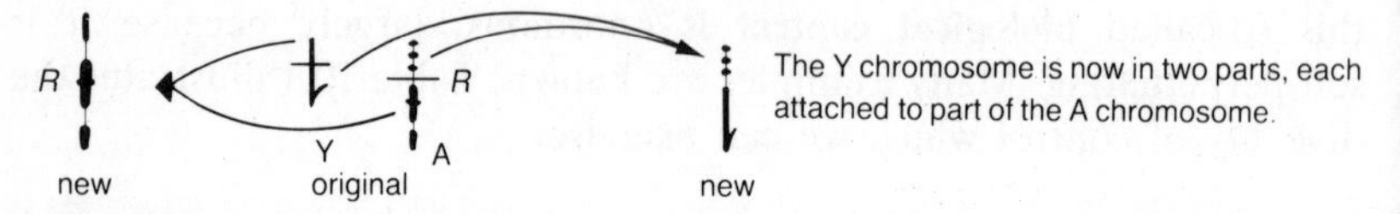

4 Although the X chromosome and the other A chromosome have no translocation, they are still involved in pairing at meiosis.
The arrangement looks like this:

All male flies have this arrangement;
there are no chiasmata.

5 The sperm produced by this fly will be like this:

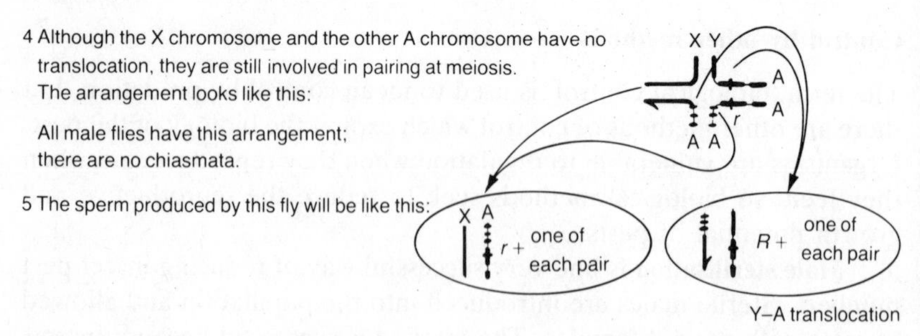

6 When this sperm unites with normal eggs from normal susceptible females (XX chromosomes and *rr* at the *R r* locus)

(a) all *male* progeny are heterozygous for *R/r*, so they are resistant (*Rr*);

(b) all *female* progeny are homozygous to *rr*, so they are susceptible.

Therefore all females can be destroyed by treating the non-mobile larvae with insecticide;
males are not affected by this treatment.

7 Sterilise the males prior to their release.

Figure 10.3 Chromosome translocations used to enable automatic sexing of insects prior to male sterilisation (courtesy of C.F. Curtis).

we could not control a mammal pest this way because the female would simply remain sexually receptive until pregnant.

Screw-worms are an example of a pest which has largely been eliminated through male sterilisation. In this devastating disease, eggs are laid in wounds of mammals and the larvae feed and grow in the living tissue of the host; death of the host is quite rapid. Cattle were affected seriously in the USA; losses to food production in 1957 were estimated to have cost the country $30m.

The male insect can be sterilised using ionising radiation. The level of radiation, X-rays or γ-rays, has to be carefully judged; too little yields no significant sterilisation, too much kills the animal. Radiation must be applied after the body cells have stopped dividing but before the germ cells have produced sperm; a pupal stage is ideal. Chemical sterilants may also be used.

The principle underlying sterilisation is that whereas the non-dividing cell can resist irradiation, chemicals or other changes in the cell's environment, rapidly dividing cells are vulnerable. So it is that these methods can also be used to check cancerous growth in humans; cancerous cells have a much shorter cell cycle than do normal cells.

The use of sterilised males in the control of screw-worm has been so successful that geneticists are developing long-term and self-perpetuating mechanisms to aid sterilisation. Figure 10.3 summarises such a scheme.

Insect lures and **hormones** are known to have a powerful effect on the behaviour of insects. Insects have a powerful sense of smell and can detect minute quantities of the chemicals released by food or by potential mates. Thus the yellow fever vector (*Aedes aegypti*) is attracted to the small quantities of lactic acid found in human sweat; and an extract from virgin female pine sawflies (*Diprion similis*) of less than 10^{-6} g attracted almost 1000 males within five minutes! The opportunity for mass eradication is enormous. Work is in progress to establish the range and nature of these powerful lures. An overriding feature of any such programme will be the cost of preparing the chemicals.

Although the enormous potential of insect hormones was heralded in the 1950s ('a teacupful would kill all the insects on one hectare of land'), their use has suffered from economic and other drawbacks. Juvenile hormone will prevent the adult emerging from the final larval moult, but it needs to be applied regularly because eggs are not laid, and larvae do not grow, synchronously. Its action is non-selective. Likewise, the hormone ecdysone, which causes moulting, will affect both useful and harmful insects. It has the further disadvantage that it is a steroid, and so may affect us and other mammals directly – several of our hormones are close relatives.

Integrated biological approaches

When controlling the spread of a disease, we are interfering with the aspects of the natural exponential growth of the disease agent. The examples given illustrate particular methods of regulation. It should be remembered that the use of only one method upsets the balance to the natural ecosystem of the food organism, its pests and the predators of the pests. To the extent that it favours the food organism at the expense of pests and diseases this is a favourable change, but it may have disadvantages also. Over-use of DDT, for example, has contaminated our food chain, has encouraged insects which became resistant to its effects, and has caused others to become pests which before the use of DDT had been kept in natural check by predators.

An example of this last problem may help. In the 1940s, DDT was used to control codling moth whose larvae bore into apples and reduce their sale value and storing qualities. The codling moth was indeed eliminated – but so were the predators of the red spider mite. This mite had been of little importance, but then assumed major significance by damaging the photosynthetically active leaves of the apple trees.

Our knowledge of biology can be used to guard against these unfortunate events and to develop a pest management or **integrated control** policy in which several methods – including hygiene, ecological and narrow-spectrum chemical control – are used. Though the phrase has become fashionable only recently, the practice has been with us for a long time. It represents good husbandry – biology in the service of humans.

An integrated approach has been used for some time in the growing of sugar beet in Britain. It is centred on the control of virus yellows, a virus disease which reduces the yield by up to twenty per cent and which is transmitted by an aphid (*Myzus persicae*). The virus inoculum originates from plants which survive the winter.

Some of the overwintering plants are those grown for seed, and these are grown a long way from those used for sugar. This also reduces the likelihood of a nearby population of aphids, because the aphids also overwinter on beet. Other members of the beet family such as mangolds, which harbour both virus and aphid, are not stored through the winter or grown close to sugar beet fields. During the growing season a spraying programme is followed if more than one aphid per four plants is observed. This uses a low-persistence, narrow-range, systemic insecticide which kills the aphids as they suck the sap of the plant. The whole management package controls virus yellows in sugar beet effectively.

Ecological control is rarely effective in isolation from other control measures. The great commercial success of the use of *Encarsia* (Figure

Figure 10.4 Whitefly parasites (courtesy of the Glasshouse Crops Research Institute).

Figure 10.5 Whitefly, and the black scales formed when the immature scale-like nymph is parasitised (courtesy of the Glasshouse Crops Research Institute).

10.4) in regulating numbers of whitefly (Figure 10.5) in tomato glasshouses is tightly dovetailed with other activities. Vulnerable young seedlings must be protected in the nursery and during transport to the growing glasshouses; chemical sprays or granules in the potting compost are used. Some whitefly must be introduced with the *Encarsia*, because the parasite itself needs a starting quantity of food. We do not have to worry about introduction of male *Encarsia*; reproduction is parthenogenetic. The rate of increase of predator numbers is affected by temperature. At temperatures above 22 °C, the parasite grows faster than the host; at 18 °C, the host grows five times as fast as the parasite; below 13 °C, the parasite quickly dies. This means that at low temperatures, chemicals must be used. The whitefly must not be allowed to overwinter either in the glasshouse or in surrounding vegetation; we use fumigation in the house and total contact herbicides such as Paraquat (page 58) on plant material in and around the houses. If the whitefly population exceeds one fly in ten plants, chemical sprays must be used because the predator cannot cope with the likely explosion in whitefly numbers. The parasite, having been killed, must be reintroduced, which explains the need for insecticides which break down rapidly.

To summarise, integrated pest control is a combined management package, each component of which makes the others more effective. Low-persistence insecticides allow ecological control. Complete control prevents the development of insecticide resistance. Hygiene prevents the introduction of a large initial inoculum, so that sensitive methods of control may be successfully used. We are no longer forced round the vicious circle of pesticide usage, in which we want new and more effective chemicals because the pest is resistant to the old one.

Summary

A pathogenic disease makes less food available to us and to our animals. In trying to reduce this waste, we might be able:

a to identify the genetic control of resistance so that resistant strains may be developed or resistant species brought into cultivation;

b to identify the biological basis of the susceptibility of the pathogen to chemical control;

c to identify the biology of control of the pathogen by predation or infection.

We can then establish an integrated approach whereby a balanced, biologically based combination of these three options is used.

11 Biotechnology – food from bugs

Micro-organisms may be disease agents; Chapter 10 considered the control of the serious loss of yield which may result from infection of crops or animals. This chapter explores ways in which the enormous growth potential of micro-organisms may be put to use to provide food for us or our animals.

For several centuries, we have known how to preserve the food content of milk as cheese or yoghurt, and carbohydrate as the alcohol in beers and wines; we are now developing the technology to culture micro-organisms and use their cells as food.

Micro-organisms, like any other feeding organism, require energy and nourishment in their feeding environment. We must therefore consider whether they rely upon energy which is renewable or non-renewable: energy from sunlight or from fossil fuels. Different energy sources have different economic values. Some sources are wastes from the paper industry, potato processing, and oil refining; others are valuable materials which could, or even should, be used for other purposes.

We must consider also the acceptability of producing food using micro-organisms; some procedures require changes to the organisms' genetic constitution (genetic engineering) and there is always the danger that a harmful type may be created. Again, the food may contain poisons left over from the raw material or produced by the organisms. Do these considerations mean that foods produced in this way should be fed only to our animals, or can we ourselves eat them? Should we even enlist the services of different micro-organisms to 'purify' these foods?

Preserving food with micro-organisms

Most foods when stored are liable to some form of deterioration. Microbes may deprive us of food not only by reducing the yield from our food-producing organisms, but also by spoiling the food when produced.

Preservation of milk

Milk is particularly susceptible to deterioration. Spoilage is rapid because milk is a liquid, which allows the microbes to disperse easily through the whole of the food, rather than remain isolated as they do on the surface of a solid food. Secondly, it contains a rich mixture of food for the microbes; cow's milk contains about 3.5 per cent fat, 3.5 per cent protein and five per cent milk sugar (lactose).

The dairy industry has particularly high standards of hygiene which prolong the shelf life of milk. Farms are regularly inspected and milk tested to ensure that micro-organisms do not contaminate the milk during or after milking. Milk is collected and transported quickly to dairies without contamination; it is then heat-treated (**pasteurised**) to reduce the chances of deterioration.

Milk production often exceeds its use, especially at certain times of the year when food for cattle is abundant, and special methods have been developed to preserve its food value in a convenient form. One way of preventing spoilage is to remove the water, giving dried milk. Another is to introduce micro-organisms into milk and encourage these to grow; they are chosen to increase the acidity of the milk and so prevent *other* micro-organisms from growing in it. The method uses particular micro-organisms under precisely defined conditions, and is used in cheese and yoghurt production.

Production of cheese and yoghurt

Cheese and yoghurt are produced by introducing into milk bacteria which belong to the genera *Lactobacillus* and *Streptococcus*. These respire anaerobically, which is helpful in liquids (such as milk) where only limited quantities of oxygen can gain access across the relatively small surface area presented to the air. Under these conditions, fermentation occurs; lactic acid is produced from the breakdown of milk sugar by glycolysis. Some of the nutrients in milk are thus used to produce acid conditions; the acid conditions in turn prevent, or at least inhibit, the growth of food-spoiling micro-organisms. (The benefits of high lactic acid levels produced postmortem in meat were discussed on page 125.)

The organisms which produce yoghurt are *Lactobacillus bulgaricus* and *Streptococcus thermophilus*; they are added as a small 'starter' culture to the milk and incubated for only a few hours at about 45 °C. They reduce the pH to 4.6, at which acidity the milk proteins are precipitated; this accounts for the watery curds which characterise certain yoghurts. Yoghurt has a shelf life of some three weeks, as compared with the few days only for milk itself.

Cheese production is very similar. By selecting a particular species of *Streptococcus*, the initial incubation may be carried out over a range of temperatures, the highest being almost 50 °C.

As in yoghurt production, the pH is lowered by the formation of lactic acid, the growth of other micro-organisms is inhibited and the protein is precipitated. Water is removed from the curded milk as whey, and the solid cheese matured.

Maturation allows further anaerobic fermentation to take place rather more slowly, and its effects are more variable. The products of this secondary fermentation impart characteristic flavours to cheese. Special conditions may be needed for maturation. Sometimes microbes may be added after the curd has been made to produce a cheese of specific flavour. *Penicillium roqueforti*, for example, is added to a certain sheep milk cheese to make the distinctive Roquefort; cheese may be curded at farms over a wide area, but must be brought to Roquefort for fungal maturation. A cheese such as Cheddar can be made throughout the world because curding is induced chemically and there is no fungal maturation; *biological* curding is generally carried out regionally. Curding by either method is followed by a period of maturation.

Thus quite simple procedures enable us to store milk in a highly nutritious form, although there is considerable waste if whey is discarded. These techniques reduce normal spoilage so that surplus liquid milk can be preserved as dairy products. They bridge the seasonal troughs and gluts of milk production.

Farmers have adopted the same principle of lowering the pH of perishable foods in making silage from grass, maize and other plant products for use as winter cattle feed.

Production of ethanol

Microbial action can also produce ethanol from certain carbohydrates, so that the energy value of the original carbohydrate can be stored. The substrate (the substance broken down) is again sugar, and again anaerobic respiration by glycolysis is employed, but ethanol, rather than lactic acid, is formed. The ethanol protects the food value against spoilage. We can see just how much energy is stored by comparing the energy content of sugar and that of ethanol (Figure 11.1).

This may seem a strange method of preserving carbohydrate, yet an area may produce, say, grapes very efficiently, while a cereal such as

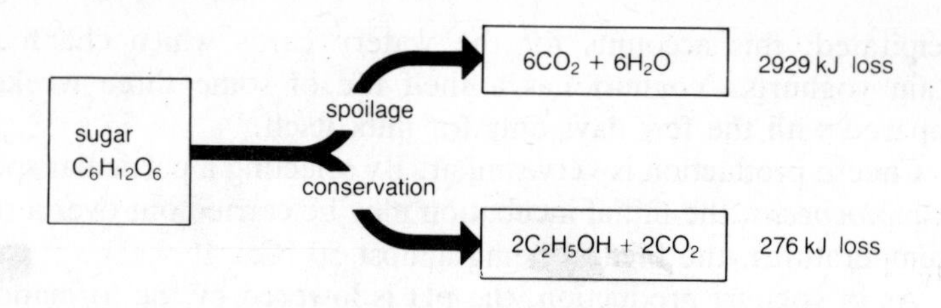

Figure 11.1 A scheme showing that in the fermentation process, the energy of the sugar is conserved as ethanol.

wheat grown in the same area gives very poor yields and the grain may require elaborate storage to prevent spoilage. By contrast, the grape juice, once fermented, keeps well in sealed casks or bottles. Alcohol thus provides certain groups of European farmers with up to one-third of their energy intake during the winter and this is significant in a peasant economy. Even in the UK, alcohol provides significant energy, six per cent on average. We now produce alcohol from material which could easily be stored and could more usefully be fed either to humans or to animals, so in developed communities it now tends to be made for social reasons rather than for nutritional convenience.

The micro-organism which converts the juice of grapes into wine is a yeast (*Saccharomyces*). Yeast forms a variety of beverages by breaking down sugars derived from the carbohydrates of grains and other plant storage organs. Thus we get beer and whisky from barley, gin and vodka from potatoes, saki from rice, toddy from coconut sap (which can be distilled to give arrack) and cider from apple juice. The production of beer and whisky illustrates the principles of brewing.

Carbohydrate in cereals is stored in the endosperm of the grain as starch (a polysaccharide). Before *Saccharomyces* can ferment the starch, this must be broken down (hydrolysed) by enzymes to form a sugar, glucose, which is the only source of cellular energy (via glycolysis). In the production of beer, the starches are broken by germination enzymes produced by the grain itself. In the production of grain whisky, on the other hand, enzymes from barley are added to break down most of the starch.

The protein content of the barley must be low to avoid cloudiness in the beer. But some of the grain's protein accounts for the enzyme β-amylase, and the seed synthesises α-amylase at germination. These two, in combination, break down starches to give maltose (a disaccharide). The strength of this action is known as the **diastasic power**

(DP); even low-protein grains used in beer-making have sufficient DP to break down all the starch reserves in the seeds.

Traditionally, the seeds are soaked and allowed to germinate on the malting floors for about two weeks at 20 °C. Kilning then stops germination; drying grain prevents further enzyme action. The timing of kilning is a balance between a short germination period, in which the starch may not be fully hydrolysed, and a longer germination period, in which much of the starch may be lost through respiration. The extent of the drying and roasting during kilning affects the colour and flavour of the malt and so the beer. The grains are then ground and mashed in water, and this rehydration may allow completion of the diastatic breakdown. The skills of the brewer are now exercised in using the sweet liquid, certain yeasts, sugars and hops to produce the desired beer. Fermentation may be stopped by pasteurisation (which may extend the shelf life of the beer) or allowed to decline naturally in the cask or bottle.

With grain whisky, the bulk of the alcohol comes from maize, but it is still enzymes from malted barley which are used to break down the carbohydrate to sugars. Barleys are needed with a high enzyme content and hence high DP. Malt whisky follows the process used in beer making; there is no added carbohydrate, and low-protein barleys are preferred.

Thereafter, the alcohol content is increased and the flavouring maintained by distillation (fermentation will naturally decline once the alcohol concentration has reached a certain level, because the alcohol itself is toxic; distillation, which concentrates the alcohol, is therefore necessary to produce strong drinks).

Products are blended and stored. No longer a cottage industry, these processes use high technology; and tight security is needed for this very valuable and highly taxed liquor.

Food from single cells

We now turn from processes which developed from the desire to preserve food or energy to new techniques for *creating* specific foods under precisely defined conditions. Micro-organisms contain food substances such as proteins and oils. It is possible to grow them on low-value or even waste materials, harvest the cells and (after suitable purification and processing) use the result as food for ourselves or livestock. When the product is protein, it is called **single-cell protein (SCP)**.

Table 11.1 Protein produced per day per 1000 kg of biomass of different organisms.

Organism	Extra protein produced (kg/day)
bullock	1
soya bean	100
yeast	100 000
bacteria	100 000 000 000 000

Growth rate

Consider for a moment this fact. Given unlimited space and no nutritional restrictions, a single microbial cell dividing every twenty minutes could produce 2.2×10^{43} cells in two days; although each cell has a weight of only 10^{-12} g, the total biomass produced in these two days would equal 2.2×10^{25} tonnes – about 4000 times the weight of the earth.

This emphasises the general points made in Chapter 6 about reproductive capacity in relation to food production. In comparison with conventional food-producing organisms (Table 11.1), the attraction of this method of protein production is irresistible – or is it?

Types of micro-organism

Micro-organisms require supplies of energy and nourishment, as we ourselves do. However, they are much less demanding; nourishment may usually be supplied as inorganic minerals, since there is rarely a requirement for complex organic molecules as building blocks. The main nutritional requirement is for energy. As with conventional providers of food, this may be supplied by sunlight (to **autotrophic** organisms) or as organic molecules (to **heterotrophic** organisms). Table 11.2 summarises the range of energy sources and the organisms best suited to exploit them. Organisms suitable for single-cell protein production include autotrophic algae, heterotrophic yeasts and other fungi and bacteria.

Substrates suitable for micro-organisms

Micro-organisms are particularly attractive as a food source by virtue of the substrates upon which they can be grown; these are summarised in Table 11.2. They include compounds which we ourselves discard and which have little value as waste. Most large-scale processes dealing with organic material produce some waste – due either to inefficiency of extraction, or to a purifying stage at which some material is discarded. Here are some examples of how these 'wastes' can be used.

Table 11.2 Energy sources available to micro-organisms, with examples of organisms best suited to exploit them. **a** renewable resources, **b** non-renewable resources.

a

Energy source (renewable)	Upgrading of substrate	Organism
cheap		
↑ CO_2 + light	no	algae (e.g. *Spirulina maxima*), photosynthetic bacteria (e.g. *Rhodopseudomonas*)
agricultural waste (straw)	yes	celluloytic bacteria (e.g. *Cellulomonas*)
sulphite liquor from paper industry	yes	yeasts (e.g. *Candida utilis*)
animal faeces	yes	
high-quality waste		
starches from food processing (e.g. potato)	no	yeasts (e.g. *Endomycopsis* and *Candida* grown symbiotically)[1]
whey from cheese production	no	yeasts (e.g. *Kluyveromyces fragilis*)
↓ molasses from sugar production	no	yeasts (e.g. *Saccharomyces cerevisiae*)
expensive		

[1] Symba process (page 174)

b

Energy source (non-renewable)	Upgrading of substrate	Organism
cheap		
↑ straight-chain alkanes (C_{10}–C_{23}) (wax waste products from oil distillation)	yes	yeasts (e.g. *Candida lipolytica*)[2]
methane (waste from oil fields)	yes	bacteria (e.g. *Pseudomonas methanica*)
↓ methanol (from methane)	yes	bacteria (e.g. *Methylophilus methylotrophus*)[3]
expensive		

[2] BP production of 'Toprina G' (page 174)
[3] ICI production of 'Pruteen' (page 171)

The cheese-making industry in the USA produces more than five million tonnes of whey each year, of which about four million tonnes are discarded. This represents about 36 000 tonnes of protein and 200 000 tonnes of lactose. Some yeasts can grow on this waste product and the harvested cells can be used as cattle feed supplements; they contain high-grade protein and vitamins (thiamin, riboflavin and ascorbic acid).

Similarly, molasses (a product of sugar refining) constitute four per cent of the sugar harvested, a loss amounting to 900 000 tonnes each year, worldwide. Some is fermented and distilled to make rum, some is fed directly to cattle (page 191), but yeast yields from molasses are particularly significant (400 kg of molasses yield 100 kg of dried yeast).

The wood-pulping industry produces a waste liquid known as sulphite liquor containing up to three per cent sugar which can support fungal growth. But the industry illustrates a general point; like many industrial processes, it is undergoing continual modification. The sulphite method of paper production is gradually being replaced by the Kraft process, and so production of sulphite liquor is declining. Therefore, it is important to consider the *supply* of the waste product.

The examples so far involve renewable energy sources for microbial growth; ultimately the energy derives from the sun. Micro-organisms can also utilise energy from carbon in non-renewable energy sources loosely associated with the oil industry.

It is estimated that during oil recovery, 100 million tonnes of natural gas are burnt off annually. If economic considerations permitted, this gas could be liquified under pressure and later used as a conventional fossil fuel. It could also be converted by chemical means to methanol, possibly at the wellhead. Imperial Chemical Industries (UK) are considering the idea that methanol, which can be transported easily, could be used economically as an energy source for bacterial growth, and so produce a high-value protein.

The wax-like products (straight-chain alkanes) of oil distillation can also support growth of yeast which can be fed subsequently to animals. Hitherto these waxes have been considered low-value products, as they need to be modified chemically to be useful elsewhere in industry.

The food potential of single-cell protein

The development of SCP is not necessarily the way of the future. It can certainly be used to upgrade wastes. It is economically viable, judging from the substantial investments by large chemical companies throughout the world. But several questions regarding the biological use of these foods need to be answered. Are they nutritious? Do they contain any toxins? Who or what eats them? We shall consider some of these.

Acceptability of SCP

The harvesting and processing of SCP yield a flour or meal, and it is the acceptability of this which we must consider.

Algal meal or freshly harvested 'soups' can be fed to humans at fairly high levels (30–40 g protein/day), and aid good health; these products could therefore be used as supplements to a varied diet.

Similarly, yeast-based supplements can be eaten without problems. Bacteria-based supplements, on the other hand, may result in nausea, vomiting and diarrhoea in humans; these are likely to be fed only to non-susceptible animals, and only in carefully controlled quantities.

Contaminants of SCP

Some SCPs are likely to contain contaminants from the substrate on which they are grown. So-called 'petro-proteins' are banned in Japan and their continued production in Europe is under review; they include oil products which may be carcinogenic (cancer-producing). Extraction of these contaminants may be incomplete; total purification will remain prohibitively expensive.

We must also concern ourselves with the changes in the genetic make-up of the organism which produces the SCP; this could become a pathogen. *Pseudomonas* (which can be grown on methane, other hydrocarbons, or methanol) has close relatives which are pathogens. That bacterial DNA is susceptible to change is shown frequently when microbes become drug-resistant. As some micro-organisms might be a potential source of disease, scientists believe that each SCP process should contain a stage when there is total denaturation of the organism, and that the culture of these organisms should be under suitably tight security.

Fears concerning the inclusion of harmful compounds from the substrate have led microbiologists to lengthen the food chain. The contaminated micro-organism-based food is used to grow *another* micro-organism and this *second* feeder is used as a supplement in our food. Chapter 2 argued that a long food chain demands a lot of energy to support it, so fewer people benefit than would benefit from the shorter chain.

An alternative approach is to select an organism which yields a specific product and to grow this on the harmful waste; this product may be purified and used to grow micro-organisms for SCP. This principle has already been applied in relation to sewage; micro-organisms which produce methane can be grown on sewage, the methane can be purified and totally divorced from the actual sewage, and other micro-organisms grown on the methane to produce SCP which is acceptable as a food additive.

Nucleic acid in SCP

SCP is likely to contain the nucleic acid (DNA and RNA) of the micro-organism. Nucleic acid forms a higher proportion of dry weight (fifteen per cent) in micro-organisms than in, say, meat or fish (liver has four per cent, for example). Disorders such as gout, once thought to be acquired only by extravagant living, are now known to result from high levels of dietary nucleic acid. Nucleic acid is broken down to uric acid, and this is relatively insoluble; it may produce stone-like accretions which block the flow of urine in the kidney and bladder. Nucleic acid may be removed from food either by hydrolysis with acid or alkali or by treatment with an enzyme; both methods are costly.

Again we see that SCP is unlikely to provide a total substitute for conventionally produced protein, and that its use will be restricted to supplementing our diets or those of our farm animals.

Amino acids in SCP

To be of nutritional value, SCP must have quantities and proportions of amino acids comparable with or superior to cheaply produced plant proteins.

We have seen that the proportion of essential amino acids is important in determining how our bodies use protein (page 12), although this was not important for the ruminant (page 29). The amino acid content of SCP justifies the name 'protein' (Table 11.3) and the composition compares favourably with fish meal. The proportion of *essential* amino acids would favour the use of SCP on strictly nutritional grounds (Table 11.4).

Table 11.3 Protein content of micro-organisms, plants and animals.

Source	Protein content (% dry weight)
soya flour	47
yeast	48
egg	49
fish	55
cow muscle	57
algae	59
bacteria	63

With certain reservations, then, SCP can provide a rich protein food supplement of high quality which will feature significantly in future food production. SCP is unlikely to provide a total substitute for conventionally produced protein, and its use will probably be restricted to supplementing our diets and those of our farm animals.

Table 11.4 Amino acid composition of high-protein feedstuffs (g per 100 g of product).

Amino acid	Toprina G	Pruteen	Fishmeal	Soya bean meal
lysine	4.2	4.3	4.9	2.8
methionine and cystine	1.8	2.4	2.6	1.3
arginine	2.9	3.7	3.6	3.2
histidine	1.2	1.5	1.5	1.1
isoleucine	2.7	3.0	3.2	2.2
leucine	4.2	4.9	5.0	3.4
phenylalanine and tyrosine	4.7	4.8	5.2	3.8
threonine	2.9	3.3	3.0	1.9
tryptophan	0.84	0.62	0.86	0.6
valine	3.2	3.8	3.7	2.3

Examples of single-cell protein

SCP from methanol (using a bacterium)

Methanol is manufactured from carbon monoxide and hydrogen, themselves formed when methane, steam and a little oxygen react under the right conditions. Coal and natural gas may be raw materials. The bacterium *Methylophilus methylotrophus* grows well on methanol as the sole source of carbon and energy, and its cells are rich in protein. Methanol proves suitable as a substrate because it is miscible with water, and the bacterium can convert it efficiently to bacterial protein.

Scientists at Imperial Chemical Industries combined their skill in manufacturing methanol with the properties of this particular bacterium. They found that a suitable way to grow these bacteria is as shown in Figure 11.2. Bacteria, methanol, and gaseous ammonia (as the only source of nitrogen) are fed into the fermenter. The ammonia is forced into solution because of the high pressure created by the column. As the whole mixture rises in the column, the bacteria grow quickly, encouraged by a temperature of 40 °C. Fermentation mixture is removed continually at the top of the column; the speed of bacterial growth is matched by the rate of removal of cells and the rate of supply of new substrate.

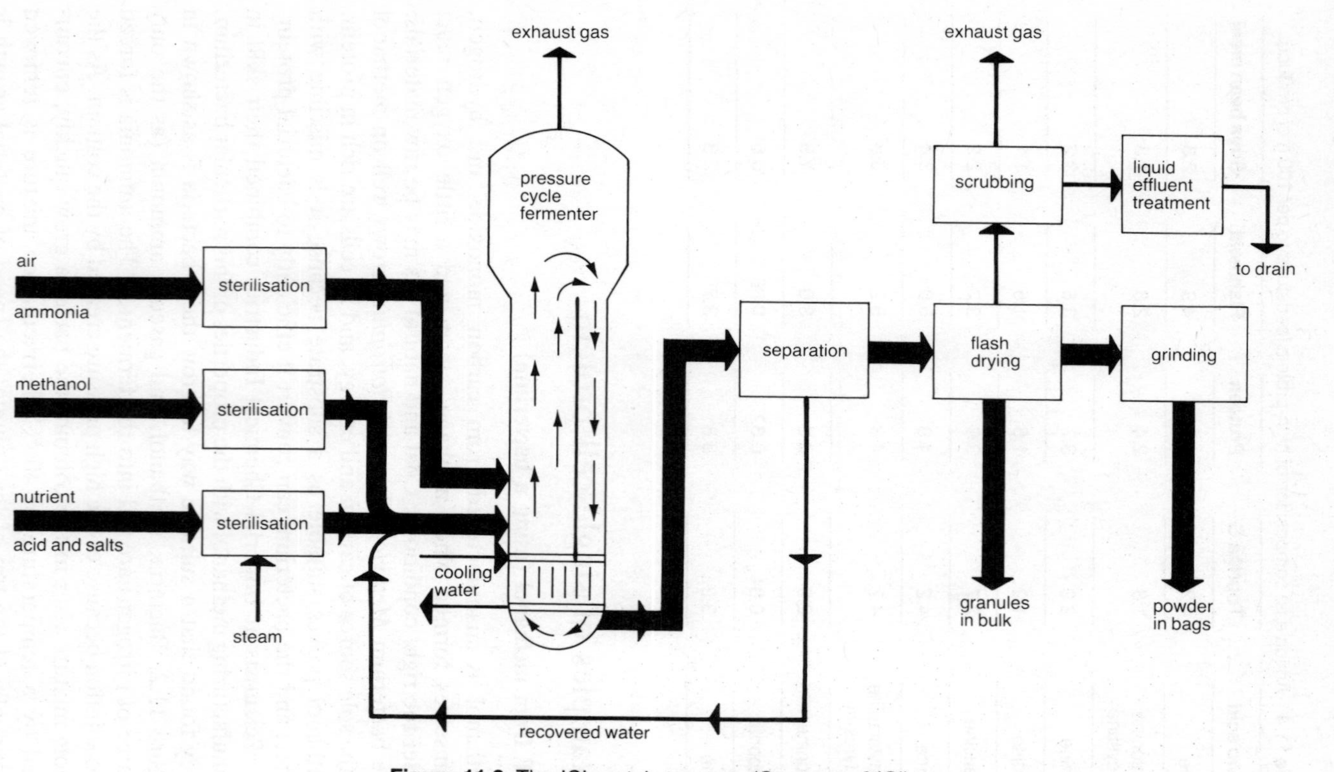

Figure 11.2 The ICI protein process (Courtesy of ICI).

The fermentation mixture is separated into cells and fluid. An increase in acidity and a heat shock curdles the fluid; the cell mass can then be filtered off. The fluid is then re-introduced at the bottom of the column. The cells are dried to prevent spoilage and the marketable product ('Pruteen') is used to supplement animal feeds. A fermenter of the type shown in Figure 11.3 may produce 50–70 000 tonnes of 'Pruteen' annually.

Figure 11.3 The installation at Billingham of the pressure fermenter. This was successfully commissioned early in 1980, and now has an output of 50–70 000 tonnes each year. (Courtesy of Imperial Chemical Industries Ltd.)

SCP from straight-chain alkanes (using a yeast)

The yeast *Candida lipolytica* grows well in straight-chain alkanes of the C_{10}–C_{23} range. Fungi do not grow as quickly as bacteria, but have two advantages; they have a lower nucleic acid content, and when fed to humans, they do not produce nausea.

British Petroleum (BP) devised a technique to produce 'Toprina G', a protein additive. The straight-chain alkane distillation products of crude oil were mixed in a fermentation vessel with water, mineral salts and air. Alkanes are not miscible with water so the whole mixture was continually agitated. Culture fluid and cells were removed continually from the culture vessel. Cells were separated from the fluid by centrifugation and then dried. The technique has not yet been found to be commercially viable, and in 1978 BP discontinued the production of 'Toprina G'.

SCP from plant carbohydrate waste (using two fungi)

A method of converting waste starch from potato processing to SCP was developed by the Swedish Sugar Corporation. This involved growth of two fungi. One, *Endomycopsis fibuliger*, produces amylase, an enzyme which converts starch to sugar. The other, *Candida utilis*, grows on the sugar. *C. utilis* is so efficient at utilising the sugar that its cells make up the bulk (90 per cent) of the micro-organism mass. This carefully controlled process requires high capital investment and high running costs. Nevertheless, the high technology of the process means that expensive acid hydrolysis of the starch to sugar is not needed, and just one person is needed to operate a continual (rather than batch) treatment of 20 m^3 of low solid waste per hour which produces 300 kg of yeast.

SCP from fruit waste (using a fungus)

Tate and Lyle, the sugar-refining company, have developed methods of utilising the waste of fruit-processing industries. Waste may be up to 70 per cent of the harvested fruit and is only a low-value animal foodstuff.

An experimental plant in Belize investigated the economics and practicability of upgrading the food value of this waste by encouraging micro-organisms to grow on it. The fruit residues are first minced by hand or simple powered mincers (Figure 11.4). An ammonium salt and minerals are then added, and the mixture introduced into a fermenter fitted with a stirrer and air compressor. Water is added to give a fermentable carbohydrate concentration of four per cent. Fermentation

Figure 11.4 The upgrading of fruit waste in Belize, using technology developed by Tate and Lyle: left, the mincing of residues; right, the fermenter plant.
(Courtesy of Tate and Lyle)

is complete after about twenty hours, when 90 per cent of the batch is removed and dried. The remaining ten per cent serves as an inoculum for the next batch. The dried product may contain up to 40 per cent protein and low quantities of energy have been used in its preparation.

Alternatively, in tropical regions, minced fruit waste may be spread out on trays, inoculated with *Rhizopus* species and then incubated in a high humidity chamber. However, because the conditions of fermentation are not precisely defined, spoilage can occur.

Fermentation of solid material is attractive as a low-technology principle, because it is not necessary to remove much water from the upgraded product. These low-technology methods of upgrading wastes into high-value, protein-rich animal food supplements do not require high investment in plant or high running costs; therefore, SCP is not the preserve of highly developed countries.

Future uses of micro-organisms in food production

Micro-organisms have a remarkable ability to adapt to meet very specific conditions in the feeding environment. There are micro-organisms which live in aviation fuel, others in waterlogged soils and

Table 11.5 Percentages of fatty acids found in oils produced by conventional and microbial organisms (based on Ratledge).

Organism	Palmitic 16:0	Stearic 18:0	Arachidic 20:0	Palmitoleic 16:1	Oleic 18:1	Linoleic 18:2	Linolenic 18:3
Yeasts							
Candida 107	36	14	4	1	36	8	–
Cryptococcus	18	6	–	1	60	12	2
Rhodotorula	31	9	–	–	53	1	5
Other fungi							
Aspergillus nidulans	18	12	1	4	43	21	–
Aspergillus ochraceus	38	–	–	–	15	45	2
Aspergillus terreus	23	–	–	–	14	40	21
Mucor circinelloides	20	25	–	2	40	4	6
Mucor mucedo	17	11	–	1	31	32	6
Penicillium spinulosum	18	12	1	4	43	21	–
Higher plants							
olive	12	–	–	–	76	6	–
peanut	10	–	–	–	58	22	–
cocoa	26	34	–	–	37	–	–
maize	8	4	–	–	46	42	–

still others in the rumens of cows. They do this by producing powerful enzymes which break down the substrate; we can use these enzymes. They also store chemicals other than those needed in their own growth; we can use these chemicals. Here are some examples of the continuing potential of micro-organisms in food production.

Single-cell oil

Some micro-organisms produce large quantities of fats and oils: **single-cell oil (SCO)**. These organisms grow on renewable or non-renewable energy sources (although methane and methanol cannot be converted into oils). Because the production of oil in this way is so attractive, attention has centred on growing yeasts and other fungi on renewable carbohydrate wastes (such as molasses, sulphite liquor and starches) which are already in use for the production of SCP.

SCO has considerable advantages over SCP. Firstly, the oil is very valuable; in 1981, the prices of vegetable oils ranged from £620 per tonne for corn oil to £1250 per tonne for olive oil, and vegetable oils account for four fifths of total (plant and animal) food oil production. Secondly, oil produced within the country which uses it saves that country money – the cost of importing the oil, a saving which improves the balance of payments. Thirdly, oils produced by micro-organisms have a constant and defined composition.

A fat is defined by the fatty acid attached to the glycerol molecule. Fatty acids vary in the number of carbon atoms and in the number of double bonds between them. Any fatty acid can therefore be represented by these two numbers, though of course we also have common names for them. (Linoleic acid, for example, with eighteen carbon atoms and two double bonds, is represented by 18:2.)

The use of a particular oil varies according to the fatty acid composition. If we know which micro-organism produces which oils, we can synthesise certain oils by growing the appropriate strain. Table 11.5 summarises the range of fatty acid compositions in single-cell and conventional oils. The production of SCO yields SCP as one of the by-products, so there is not a question of SCP versus SCO, but rather one of balance.

When these highly selected organisms are grown on specific culture media in artificial vessels, the products are predictable. The products of micro-organisms growing in the gut of the ruminant are also predictable. Diet affects the fermentation products available for assimilation by the cow; the proportion of butanoic acid (a volatile fatty acid) in the blood leaving the rumen increases progressively when the diet changes from hay, to silage, to hay plus food concentrates, to silage plus food concentrates. We can change the composition of butter: soya-based foodstuffs increase the proportion of C_{18} unsaturated fats in the milk, forming soft butter; palm-oil-based concentrates stimulate the production of C_{16} saturated butter fats, and the butter is hard.

Are there drawbacks to producing SCO? There are some, mainly technological and economic; but given time, these will be resolved. However, if we consider the efficiency of producing SCO, in terms of the output as a proportion of the input (as we did on page 31 with the conversion of food by poultry to eggs), we find that it rarely exceeds twenty per cent (20 g fat/100 g substrate), and may be as low as fifteen per cent–which compares unfavourably with the 25 per cent efficiency of SCP production.

Enzymes from micro-organisms

In some processes, enzymes may be used in place of inorganic catalysts. Glucose syrups, for example, which are used as sweeteners in confectionery, have for many years been produced from starch by acid hydrolysis. Tate and Lyle now have a factory in the UK which uses enzymes from bacteria and starch from potatoes as the raw material. A particular variety of potatoes (which often has to be imported) provides good yields of the desired starch. A two-stage hydrolysis then takes place: *Bacillus subtilis* is used to provide bacterial α-amylase which hydrolyses the α-1,4 bonds in the starch molecule to produce dextrins; the branching bonds (α-1,6 linkages) are then hydrolysed by an enzyme from *Aspergillus niger*.

The enzymes are linked to a supporting structure so that the process can be continuous as the starch flows over sheets of enzymes; this yields a cleaner product than would be obtained from starch and enzymes in an incubation vessel. Enzyme sheets prepared in this way function for up to three months of continual use – confirmation that the enzymes are not affected by the reaction and can carry out the hydrolysis many times. The use of specific enzymes in large-scale manufacturing industries has potential to increase, and such techniques may become commonplace in the food industry.

Other products from micro-organisms

Our need for particular organic chemicals in the food industry and in medicine is likely to increase. To illustrate this need, we consider now the production of a food additive, and in Chapter 12 that of a human hormone. Both rely heavily upon the manipulation of the bacterial genome.

Monosodium glutamate (MSG) is frequently added to dried 'convenience' goods, and is said to bring out the flavour. It is produced by growing bacteria in a vat-like culture vessel with a source of carbon for energy and growth, plus mineral salts. Liquid from the culture vessel is removed and centrifuged; the centrifuge tubes are fitted with filters which collect the bacteria but allow the liquid through. As they grow, the bacteria produce glutamic acid; this flows through the filter with the culture liquid, and it is a simple procedure to neutralise this and form monosodium glutamate.

This chapter has demonstrated the enormous growth potential of micro-organisms, their ability to grow on the most unlikely waste materials, and their production of large quantities of a range of food material which can be used in 'tailor-made' food production schemes. In Chapter 12, we speculate on the way that we can modify these qualities by changing the genetic material of micro-organisms.

Summary

Micro-organisms have rapid rates of reproduction, and as saprophytes they can exploit waste or low-value food substances as a source of energy and minerals. They use the food source to reproduce, and they produce specific by-products. To exploit micro-organisms, we need:

a to identify the surplus or the waste as appropriate substrates;

b to identify the nutritional value of the cells for use as animal or human food;

c to identify the beneficial food or chemical products;

d to match organism to substrate.

12 The Future

Food production, like all other aspects of life on earth, is subject to biological and physical constraints. Two of these in particular affect future developments. Firstly, food production requires fixed resources – minerals and elements which are neither created nor destroyed, but only concentrated or dissipated. Secondly, food production requires continued supplies of 'renewable' energy from the sun. As organisms grow, they accumulate this incident energy in their biomass, which we remove as food. Traditionally, fossil-based energy has been used for industrial and mechanical pursuits, and renewable energy has been used for food production.

At the end of this century, the world population may be double the present figure. Our ingenuity in manipulating for food the mineral resources and the flow of energy from the sun, via autotrophic and heterotrophic organisms, will influence our life style. This ingenuity may exploit two alternatives: the control of the environment of the food-producing organism, and the manipulation of existing units of genetic information into new combinations and new uses. What recent discoveries are likely to help us?

Controlling the environment

Optimisation of food for plants

Improvements to the environment of the plant are unlikely to include increased supplies of artificial light energy; conversion of light energy to chemical energy is of such low efficiency that only in exceptional circumstances would light be supplemented. Examples of such circumstances might be very high-value crops where early harvests command a selling premium, and crops grown for research.

Therefore improvements to the environments of plants are likely to be restricted to the judicious and efficient supplementation of minerals.

Supplies of minerals are limited, and only a small proportion of those added to the soil actually reach the plant. Additions can be made via the leaves, however; **foliar feeding** ensures that minerals get into the plant. Or we can grow plants without soil – a technique based on discoveries made by Julius Sachs in 1860.

It was as late as the mid-1970s that scientists at the Glasshouse Crops Research Institute (GCRI) in the UK developed the so-called **nutrient film technique** (NFT) of growing plants without soil. Soil is not needed because the plant's weight is taken by stakes and wires; a nutrient solution is circulated continually through the troughs in which the roots grow (Figure 12.1). The composition of the minerals and oxygen can therefore be carefully monitored, and maintained at the correct level. The roots are not submerged in the solution – a film, rather, of nutrient solution flows down the gulleys and is thus further aerated. This system allows the ultimate control of the root environment; minerals, pH, oxygen and temperature may be regulated. The system serves additionally as an automatic watering device; water lost by the plant is made good in the storage–aeration tank. The crop is isolated from the soil, so the whole mineral content of the solution may be determined by the grower. Soil-borne diseases, often a major

Figure 12.1 Young tomato plants growing in the nutrient film system. Nutrient solution enters through the flexible pipe at the upper end of the trough, which is semi-rigid. Two troughs are placed together between heating pipes.
(Courtesy of the Glasshouse Crops Research Institute.)

problem of intensive production, are eliminated. Disease-control chemicals may be added to the nutrient film if necessary. Using NFT, the tomato grower may expect a yield of 250 tonnes/ha. The capital cost involved in installing this system is high but the glasshouse and *its* running costs – heating – are also high. The use of NFT will depend on how these costs vary in relation to the cost of other methods of production.

NFT is not restricted to the glasshouse. Should air temperature and levels of sunlight be suitable for prolific plant growth in an area where water availability and soil fertility are unfavourable – as, for example, desert regions of the Middle East – NFT provides suitable environmental control.

Nutrient film technique shows how a knowledge of the biology of plant growth enables us to consider new options.

Improved feeding regimes for animals

What new practices may we use to improve the environment of our food animals, and so help future food production? Recent experiments concern alternative feeding regimes. One example is the use of waste – fish waste, abattoir offal in natural or pelletted form, or poultry manure – as food for fish! Poultry manure may seem particularly bizarre, but poultry are inefficient at digesting protein and much is voided in their faeces. So there is an obvious attraction in re-introducing this higher up a food chain which results in human food, rather than allowing it to decay in the soil and be used later by autotrophic plants.

Rumen fermentation causes vast changes to the amino acid composition of the animal's food, so protein which is of low value to humans may be profitably used by a ruminant. However, some nitrogen is lost from the organism – and hence from the food chain – as nitrogen gas. This loss may be reduced by encasing high-quality soluble protein in gelatine, so that it may be absorbed efficiently by the animal's true stomach.

Fermentation is accompanied by chemical reduction, so the bulk of the volatile fatty acids absorbed by the gut become fully saturated. (This is why beef lard is hard.) The incorporation of unsaturated fats into fat deposits in the ruminant's body may be encouraged by feeding fats which are *protected* from such reduction. This alleviates the nutritionist's objections to saturated animal fats as contributory to human heart diseases.

Energy utilisation is important in improving feeding regimes. **Homoiothermic** animals are so named because they maintain a constant body temperature, irrespective of the external temperature. Therefore,

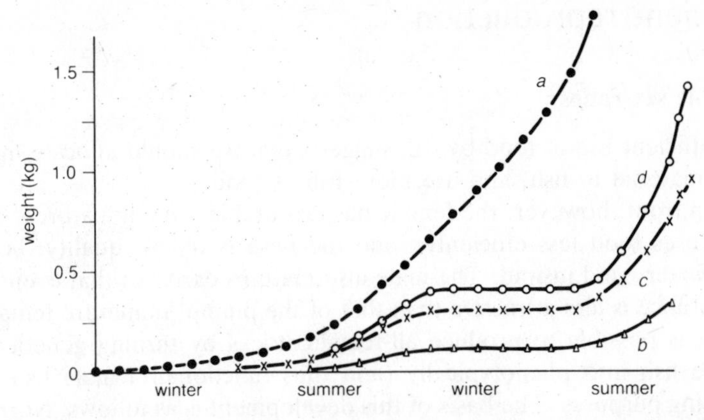

Figure 12.2 The growth of turbot reared under four different management systems: (*a*) power station; (*b*) ambient – natural surroundings; (*c*) closed-circuit water flow followed by ambient; (*d*) closed-circuit water flow followed by augmented ambient. (Based on a Ministry of Agriculture, Fisheries and Food publication.)

if their environment is cold, they use more energy to keep warm. **Poikilothermic** animals, on the other hand, have body temperatures which vary with those of their surroundings and hence use proportionally more of their energy for growth than do homoiotherms. Their basal metabolic rates reflect this; that of a placental mammal (homoiotherm) is about 3.8 m^3 O_2/(g hour), while that of a freshwater fish (poikilotherm) at 25 °C is 0.2 cm^3 O_2/(g hour). The poikilotherm, in simple terms, does not waste food keeping warm, and is thus an attractive animal to grow for food. The actual efficiency of food conversion is difficult to establish. For example, in feeding trials with fish, there is uncertainty as to whether a fish has eaten all the food in the water, but such trials do show that fish may accumulate up to one gram of fresh weight for each gram of dry food, and this suggests efficient use of food.

Because the body temperature of a poikilotherm is the same as that of the environment, the animal's growth rate may be determined by the ambient temperature. For fish, higher temperatures can be established and maintained by rearing them in water heated by waste energy from other activities. When turbot are grown in waste warm water from the cooling of a power station, higher growth rates may be maintained, especially in winter, and growth follows an uninterrupted and predicted S-shaped curve (Figure 12.2).

Efficient reproduction

Suitable sex ratios

The efficient use of food by fish suggests that we should allocate more food material to fish, and use more fish as food.

In trout, however, the female has favourable growth features. The male uses food less efficiently, and the flesh is of low quality, being soft, watery and insipid. The male also matures early, so that even the appearance is less attractive than that of the plump, immature female.

It is possible to produce all-female stocks by turning genetically female fish into phenotypically (and thus functionally) male fish for breeding purposes. The basis of this development is as follows. Normal male fish produce two types of haploid gametes; after fertilisation, one type will form female young, the other will form males. (The male is the heterogametic sex, as with humans.) If a female can be made to produce sperm, it would be of one type: that which forms female young. This would allow us to rear all female stocks.

Research shows that hormone treatment will masculinise a female; methyl testosterone – 3 mg hormone/kg food, administered over a period of 60 days at 10 °C (600 so-called **degree days**) – will cause the change. The milt (sperm) from these 'males' is all of the one type, carrying an X chromosome. A great attraction of this method of regulating sex ratios is that chemical castration is not used; the animals which we eat are not treated at all; they are reared under completely natural conditions.

We cannot yet manipulate the 'sex' of mammalian sperm, but it is an area of continuing research. Many workers have tried to separate X and Y sperm on the basis of their physical characteristics, arguing that the larger Y sperm may cause it to move differently in a gravitational field, or alter its charge so that it moves differently in an electric field. The techniques have not been successful and we now know that the genes of the sperm are not usually expressed in the phenotype of the sperm. The characteristics of a sperm are determined by the genotype of the testis! Only rarely is there haploid gene expression.

The sex of young embryos can be determined, however. At the blastocyst stage, the embryo may be removed from the uterus and a small portion used for cytological examination. The presence of a Y chromosome indicates a male embryo, and so the sex of all embryos can be established. Embryos of the appropriate sex are returned to the uterus. While this is too expensive for general use, it may be used when an animal of very great genetic value is being multiplied by superovulation and embryo transfer (discussed later).

As our knowledge of the control of the mammalian gonad increases, we may be able to progress to the solution adopted with the trout. We see again that more basic knowledge is needed if we are to increase our options for food.

Rate of reproduction

All food is dependent upon reproductive excess. The low reproductive rate of animals is therefore of particular concern to the agriculturalist in two ways: directly in the production of more food, and indirectly in the propagation of the valuable genes of the superior animal.

There are two methods of increasing reproductive potential. The first is to increase the number of offspring produced from each organism on each reproductive event; the second is to ensure that the organism is reproductively active throughout its normal life.

The number of offspring produced may be increased by **cloning**, whereby scientists encourage non-reproductive cells in superior animals to develop into new individuals. The nucleus from the cell of the animal to be cloned is transferred to the cytoplasm of a fertilised egg whose own nucleus has been removed or destroyed. Clearly, the egg must be of the same species, and – for higher vertebrates with internal development – the egg must be in the reproductive tract of an individual in a suitable condition.

By providing a suitable environment, we must persuade the nucleus from (say) the gut of an animal that it is about to undergo cleavage and early embryo development, and we have cloned the superior animal. While we can do this with lower vertebrates (such as amphibians) which develop outside the female's body, only now are there the first tentative reports of very rare success in higher vertebrates.

Although we cannot clone higher vertebrates, we can at least divide the early cleavage stages after fertilisation, and so produce identical twins, triplets or even quads. The extent of this subdivision is restricted by the number of cells and the quantity of cytoplasm remaining in the early blastula stage. Age, rather than number of nuclei, affects control of early development. Embryos form blastocysts at a specific age; if there are too few cells, they cannot do so properly. Surprisingly, even if up to 90 per cent of the nuclei are killed in the young gastrula, normal development can still take place. This kind of manipulation is still largely a research tool and we need to know much more about the control of development before we are likely to be able to delay blastocyst formation while cell division continues. Then division of the cell mass into a larger number of potential embryos could be effective in animal improvement.

We can, however, cause the superior female to **superovulate** (shed many eggs), allow these to be fertilised, and then transfer the eggs to the reproductive tracts of suitably synchronised females by the process known as **embryo transplantation**. By hormone manipulation, the female produces more reproductive cells than could develop naturally into young in her own uterus, and the embryos are grown to parturition in 'nurse' animals. The system can be further simplified by freezing the embryos and storing them in liquid nitrogen until they are required or until appropriate mothers are available.

A further aid to increasing reproductive potential is **fertilisation *in vitro***, where ova and sperm fuse in a test tube. This has been successfully carried out in humans and in laboratory animals, but not as yet in farm animals. In humans, the technique permits pregnancy when some physical deficiency (such as blocked Fallopian tubes or very dilute semen) preclude fertilisation in the usual way. In farm animals, the real advantage of fertilisation *in vitro* would be as one step towards the supply of large numbers of ova for transplantation. This would allow either the transfer of a second egg into a pregnant cow to produce twins or the rapid reproduction of superior genotypes. The technical problems of fertilisation *in vitro* are nearly solved for farm animals, but there is another problem; we need a good egg supply.

Scientists are now attempting to mature oöcytes *in vitro*. Consider the micro-organism-like potential of this; as it is, the mammalian ovary contains up to half a million oögonia, of which only a few hundred mature during the animal's natural reproductive life. All might potentially be available when we understand oögenesis.

Pregnancy testing helps to ensure that the animal's reproductive capacity is fully realised. Each cow is expensive to maintain; if its reproductive condition is understood, farmers can ensure that each animal is making milk, or milk and a calf, or just a calf, at the correct stages of her life.

The cow is normally mated some three months after she has calved, so that she will calve annually; delays are expensive. Progesterone levels in the blood rise after ovulation, but remain high only if implantation takes place. Progesterone is passed into the milk, in proportion to its concentration in the blood. In the human, large quantities of chorionic gonadotrophin, produced by the placenta, are lost in the urine. This is not so in cows, so that milk samples rather than urine samples are taken, and the quantity of progesterone is assessed by **radioimmunoassay**.

The milk under test is mixed with a known quantity of progesterone which contains a radioactive label. A small quantity of an antibody to progesterone is added and an antibody–antigen complex precipitates. The radioactivity in this precipitate is then measured. If the milk

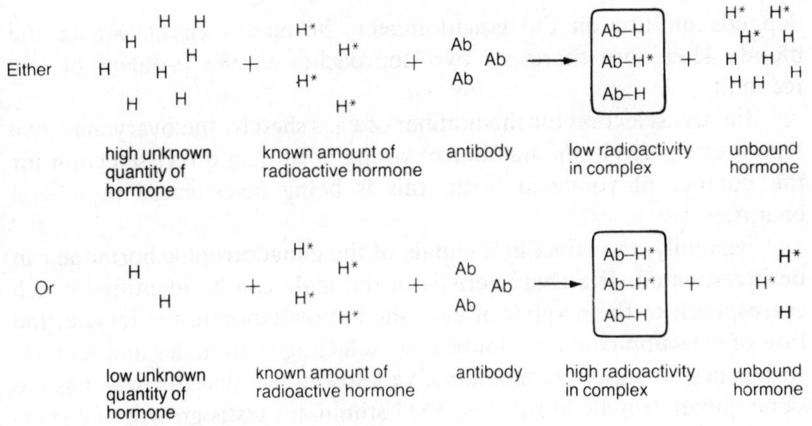

Either H H H H H H H H + H* H* H* H* + Ab Ab Ab ⟶ [Ab–H Ab–H* Ab–H] + H* H* H* H H H H H H

high unknown quantity of hormone | known amount of radioactive hormone | antibody | low radioactivity in complex | unbound hormone

Or H H + H* H* H* H* + Ab Ab Ab ⟶ [Ab–H* Ab–H* Ab–H] + H* H* H

low unknown quantity of hormone | known amount of radioactive hormone | antibody | high radioactivity in complex | unbound hormone

Figure 12.3 The principles of radioimmunoassay.

contains high levels of progesterone – as with a pregnant cow – the radioactive hormone is swamped and the complex formed has little radioactivity. Conversely, if the milk contains low levels of progesterone – as with a non-pregnant cow – the radioactivity in the precipitate is high (Figure 12.3). Thus the farmer can ensure that the cow conceives, or he can cull it.

Pregnancy testing is now in everyday use, and is an illustration of how a research technique in one decade can become a commercial tool in the next.

Efficient selection

The scientist must be alert to new ways of identifying traditionally important characteristics. A more clearly identified trait may lead to more rapid selection.

Yield is often dependent on many small factors each of which has a high heritability. It is important to recognise each factor and select for its improvement. For example, it is well established that most carbohydrate in wheat grain comes from photosynthesis by the flag leaf (the leaf just below the ear) and by the ear itself. So breeding programmes have sought to increase the area of this leaf. The flag leaf is important at the moment, but there is no certainty that its contributions will continue to increase with its area; other leaves may become important.

Small factors contribute to fecundity in mammals. The number of young born depends upon the number of eggs fertilised, which depends in turn upon the number of eggs leaving the ovary at ovulation, which depends in turn on the gonadotrophic hormones circulating in the blood. There are therefore two approaches to the problem of low fecundity.

Firstly, selection for the number of eggs shed by the ovary may give a greater change in the number of young born than direct selection for the number of young at birth; this is being investigated in several countries.

Secondly, the effect in the male of the gonadotrophic hormone can be investigated. If a characteristic in the male can be identified which corresponds to the number of eggs shed at ovulation in the female, the rate of selection could be doubled by selecting both males and females with genes for superior fecundity. We know in fact that the male has the same gonadotrophic hormones; FSH stimulates testis growth and spermatogenesis, LH stimulates growth and testosterone secretion. So if testis size reflects the gonadotrophin level in the male just as ovulation rate reflects its level in the female, we can identify males with genes which promote good female performance. This has been tested in mice; males with large testes sire daughters which shed larger numbers of eggs. Scientists can now develop this in domestic animals.

Genetic engineering

If we wish to change the genotype of an organism, two possibilities exist. Firstly, we can choose among organisms, phase out the original breed and grow for food another more favourable one in the same habitat. This is selection; it takes place either between existing populations (species or breeds), or among existing individuals of one breed.

Alternatively, we can introduce genes from another organism to enable the existing one to produce food or other derived material more efficiently. This is **genetic engineering**.

Why do we need genetic engineering? We have already seen many examples of selection and of its results – the characteristics of fish which make their growth suitable for food production, the use of NFT to allow lettuce to be grown in the desert, and the exploitation of micro-organisms as a realistic (if unconventional) method of making food from oil.

Population substitution is rarely so dramatic. Generally, the farmer is alert to the potential of new or different breeds and strains as he is alert to the features of different species, and substitution is gradual.

Selection allows us to change the balance between the different food products offered by a particular organism, and to increase the efficiency of their production. We have seen some of the dramatic effects and the enormous increases in yield of food that have been effected. It is likely to continue to be the principal source of genetic improvement in our food-producing organisms, but there are some limits. For example, it rarely allows us to produce something entirely new. We cannot separate the organism from its products, either body mass or secreted substances; we must accept these if we grow the organism. Often we would like a particular product but are unable to achieve it either with existing organisms or in existing environments.

There are some circumstances, then, when we would like an organism to produce something entirely new; this is what genetic engineering aims to achieve. It is carried out by combining specific genes at the molecular level or by combining different cells (so-called para-sexual methods) to make a suitable organism.

Human growth hormone

Normal growth of babies requires growth hormone (GH), secreted by the anterior lobe of the pituitary. Low secretion results in retardation – reduced growth or dwarfism. If low secretion is detected early enough in a baby's life, normal growth may be maintained by the regular injection of GH.

Human growth hormone (HGH) is not the same as the growth hormone of other species, in fact HGH is altogether a much smaller molecule, so we must use HGH. The major source of this is in the pituitary gland of cadavers (newly dead). This supply is limited, of course, and it is an unpleasant and difficult job to collect it.

Because of the exact replication of the nucleus at cell division, all cells in one body contain the same genetic information, but for a given cell, only part of this information is normally used – a part depending on the position of the cell in the body. So in seeking a source of HGH *in vitro*, all cells are potential candidates. Cells of the anterior pituitary, while an obvious choice, cannot be used because they will not divide *in vitro*; but the cells of foetal blood will. Even then, we cannot use these cells to produce HGH directly, because the genes which code for the production of HGH are not 'switched on' and because the cells will divide only a limited number of times in culture.

Genetic engineering permits us to incorporate the DNA code for HGH from human cells into another organism which divides and grows repeatedly with the HGH genes 'switched on'. The host organism chosen is *Escherichia coli*, and the source of HGH genes is foetal blood cells. The stages of transfer are complex.

Firstly, the mRNA which codes for the precursor of HGH must be identified and purified. Secondly, the end of the mRNA chain is digested away so that only that section which codes for the hormone itself is left. This is used as a template from which the HGH DNA is synthesised. (Thus the HGH gene is itself synthesised in the laboratory.) To function properly, the HGH gene also needs to be preceded by a promoter sequence (DNA responsible for turning the gene on and off), a ribosome binding site, and an initiator codon (the start of the DNA message), and it must be terminated by a specific sequence of DNA bases. (The whole functional message is rather like a paragraph, with introductory sentences, a particular piece of information, and a closing remark.)

This DNA is transferred into the *E. coli* cell by means of a plasmid (a short piece of motile bacterial DNA). The modified *E. coli* is then grown up to a colony and fed the promoter chemical which switches on the gene. The particular promoter sequence involves the *lac* operon which was investigated by Jacob and Monod in the late 1950s. In the same way that the addition of lactose caused the production of β-galactosidase, addition of lactose to the newly engineered organism now causes HGH to be produced in quantity, and this can be harvested later.

Human insulin can also be prepared in the same way. Because the insulin obtained this way is human, the attendant immunological problems sometimes associated with the use of bovine (cattle) or pig insulin in the treatment of diabetes are avoided.

Nitrogen fixation

It is one thing to transfer a relatively short piece of DNA which codes for the 188 amino acids of HGH, quite another to transfer a cluster of genes such as those involved with the fixation of gaseous nitrogen. Yet the combination of the nitrogen-fixing genes of a bacterium and a major crop such as wheat would be ideal.

The enzyme nitrogenase is a complex protein with a M_r of about 300 000, and comprises several sub-units with both iron and molybdenum ions. Fixation and subsequent reduction has several steps, and operates only under totally anaerobic conditions. So straight incorporation of nitrogen fixation into wheat seems unlikely. But there are two real possibilities. Firstly, we may introduce the relevant genes from, for example, the nitrogen-fixing free-living bacterium *Klebsiella pneumoniae* into some other organism with desirable growth characteristics. This has now been achieved with the hitherto non-fixing *E. coli*.

Secondly, we may try to modify the genotype (and hence the phenotype) of a major crop plant to encourage a nitrogen-fixing bacterium to develop a symbiotic relationship, as between legumes and

Rhizobium. One approach is to seek to unite, say, the productivity of wheat and the nitrogen fixation symbiosis of peas. Their breeding systems preclude sexual mating, so genetic material must be united some other way, such as by fusing the ordinary body cells – the parasexual approach to genetic engineering. To do this, the cell wall must be removed from relatively unspecialised cells by treatment with a cellulase enzyme, producing the isolated protoplast. Then the plasma membranes of two protoplasts of different species must be persuaded to join. If the nuclei fuse there is true cell fusion, and the 'hybrid' nucleus should proceed to divide by mitosis.

There is hope that this kind of somatic cell hybrid might allow a symbiotic relationship with a nitrogen-fixing bacterium and the resulting organism, while retaining the food-producing features of the crop 'parent'. At present, for example, the yields of legume grains tend to be less than half those of cereals. To date, the problem hinges upon the fusion of the cells; cells of suitable parents seem incompatible! But it is certain that over the next few years, food producers will want to maximise the output of the legume and explore the relationships between free-living and symbiotic nitrogen fixers and crop plants. Energy is needed to fix the nitrogen applied as fertiliser to intensively producing land, equivalent to that available in up to 400 litres of diesel per hectare or about 75 litres per hectare for temperate cereal production. In the future, this energy will be needed for other purposes.

Rumen flora

The main organisms to be successfully engineered so far are micro-organisms. We have seen that the problems are considerably greater when it comes to plants and farm animals. The ruminants, however, are partly secondary feeders on the micro-organisms which grow in their gut – they present an excellent opportunity for the genetic engineer. We have the chance to manipulate the micro-organism to help ruminant growth and so aid humankind.

Although micro-organisms in the rumen do break down cellulose, they do this so slowly that, for rapid growth, the ruminant must be fed readily digested carbohydrate such as starch or sugar. The enzyme which breaks down cellulose at present is cellulase-N; this enzyme will only hydrolyse the terminal glucose units in the cellulose molecule, so no matter how much enzyme is produced, hydrolysis is slow.

Another organism, *Trichoderma viride* (a mould) produces cellulase-X, an enzyme which attacks the whole cellulose molecule so that it is rapidly fragmented. If the genes for cellulase-X could be transferred to the micro-organisms in the rumen, the ruminant could digest much higher levels of cellulose in its food – and use even lower-value food than at present.

Another exciting possibility centres on the fact that cows produce more milk in response to extra growth hormone. At present, there is no cheap supply of growth hormone, although bacteria (subject to genetic engineering and grown artificially) may provide such a source. The recent introduction of the rat growth hormone gene to mice, and its expression, points to the possibility that cows could be induced to produce more of their own growth hormone. There are opportunities for hormones which might stimulate reproduction or the growth rate of lean meat.

With cellulose-X, ruminants could use cellulose as their energy source. These could then economically use some of the cellulose residues from alcohol production. Indeed, the future for the production of 'gasohol' – alcohol to replace petrol – may depend upon the economic value of the waste products of the yeast fermentation; single-cell protein is valuable, but cellulose is not. It is suggested that almost half of the grain produced by the USA may in future be used to make gasohol. This would free the USA from its dependence on oil imports; it would also save the US Government the cost ($1 billion) of purchasing surplus grain, but deprive other nations of a source of food. Therefore there are biological and political implications of genetic engineering whose effects are profound and inevitable, as well as the implications of the relatively improbable but much publicised 'super-breeds'.

A new look at existing organisms

The scientific world is regularly bombarded with speculation that a new food-producing organism has been identified among the plants and animals which already grow on our planet. To be accepted, such an organism must show a favourable economic return. It should also accord with certain biological principles. As food flows along the food chains in nature, two things are certain. Firstly, energy is lost at each trophic level, so the quantity of solar energy needed to produce a certain weight of biomass at a subsequent trophic level increases. Secondly, because all organisms ultimately die, all the energy is lost to the decomposers. These two facts mean that we should look for food to plants, or to those herbivores which live upon vegetation which we ourselves do not exploit.

Certain countries have farmed game; the game animals show trypanosome tolerance, heat tolerance and drought resistance, so that animal productivity in areas of marginal value is increased considerably

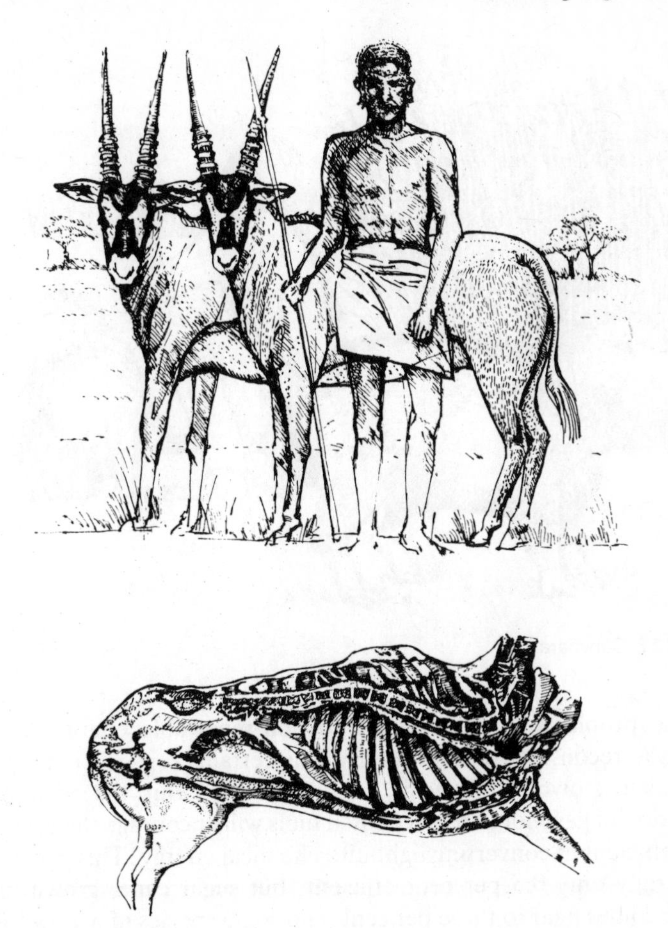

Figure 12.4 Domestic oryx, with Masai herdsman; and the carcass of an oryx slaughtered at a conventional slaughter-house.

in comparison with conventional livestock. For example, the Game Ranch Research Project, based on more than 600 000 hectares of Kenya, shows effective use of the oryx (Figure 12.4).

Similarly, the capybara is at an advanced stage of domestication (Figure 12.5). This is a large rodent like an enormous guinea pig. It grows well in the flood plain ecosystem of the American tropics and weighs about 30 kg at slaughter. It relies on post-gastric fermentation of vegetation in the caecum, its digestive system comparing favourably in efficiency with that of cow and sheep. It produces about five young per pregnancy; its 'reproductive efficiency' is six times that of cows (Table 12.1). For the same land area and length of time, capybara farming in these areas produces about 4.5 times as much food as cattle.

Figure 12.5 Capybara.

The problems of renewable and non-renewable energy reserves force us to reconsider reasons for growing certain crops. To date, plants have been grown for specific food products. But the widespread production of gasohol to replace fossil fuels will encourage the growth of plants efficient at converting light into chemical energy. This conversion is generally only 0.5 per cent efficient, but sugar cane, grown in the tropics, comes near to three per cent, and some species of willow (*Salix*) use six per cent of incident energy.

Table 12.1 Reproductive efficiency of capybaras and cattle.

	Capybaras	Cattle
a length of gestation (days)	147	275
b progeny per litter	4.73	1.0
c litters per year	1.83	0.5
d weight of dam (kg)	45	350
e average weight of progeny (kg)	1.3	28
reproductive efficiency (weight of progeny per kg live weight of dam per year): $\frac{b \times c \times e}{d}$	0.25	0.04

Conversely, we have never imagined that plants could actually produce a petrol-like chemical, but some spurges (members of the Euphorbiaceae) have a resin-like sap in their tissues which can be converted into a petrol substitute. One hectare of *Euphorbia tirucalli* may produce up to 125 barrels of oil each year, and it can grow with very little attention in arid regions.

The chlorophyll molecule, a part of plants' photosynthetic apparatus, splits water into hydrogen and oxygen. Hydrogen can be burned in a variety of ways to produce energy. A system is being developed which can produce hydrogen from water using a synthetic chemical based on chlorophyll, and so exploit the sun's energy to provide a fuel.

Perhaps the most important aspect of using efficient but hitherto unused food-producing organisms is the attitude of the food-buying public. To the biologist it seems strange that brown eggs are favoured in the UK, but white eggs in the USA. Similarly, the consumer insists on white navy beans in the preparation of baked beans, but could not brown or flecked beans (*Phaseolus vulgaris*) be used just as acceptably? White beans, 80 000 tonnes per year, have to be imported, yet many areas of the UK could grow non-white varieties. Such home production would have a dramatic effect on our balance of payments.

Problems facing the plant and animal producer

Food producers will face various problems over the next few decades. The largest will be that of responding rapidly and effectively to a change in demand for their products.

We have noted changes in the fat content of the pig carcass; here the public demand changed. We have seen the changing patterns of agriculture determined by governmental policy, and how these have resulted in food 'mountains' in the EEC. At one time there is a premium to be paid by dairies for milk with high butter fat; at another, this premium is removed. As market demand swings, the objectives of the plant and animal producer must change also, and the agricultural scientist must help him to achieve these.

Changes in climate may also have an important effect on the future. Take, for instance, the higher carbon dioxide levels created by consuming fossil fuels; the carbon dioxide in the atmosphere allows solar radiation in but does not allow the radiation emitted by structures on the earth (of greater wavelength) to escape into space. This possible **greenhouse effect** may cause a gradual heating up of the earth, which may pose major problems to our future food producer.

Perhaps the most important changing factor in agricultural practice is the availability and cost of oil. In the future, much of humankind's ingenuity will be devoted to ways of living with less oil; and the agriculturalist will have an important role to play.

Summary

To feed the ever-expanding world population, biologists have onerous responsibilities in increasing food output. Lines of investigation over the next few decades might include:

a improving the conditions under which food-producing organisms are grown so as to maximise existing resources;

b improving our accuracy in defining yield in biological terms;

c improving rates of reproduction (current levels require us to keep large numbers of breeding, rather than food-producing, individuals);

d enlisting new and hitherto unused organisms to produce new materials in unused parts of our ecosystem;

e using non-classical methods of manipulating genetic material (i) by fusing *body cells* from different species in parasexual hybridisation, or (ii) by fusing *genetic molecules* from different species so that genes from one are expressed in the other.

Taking the topics further

The subject of this book does not lend itself to practical work in schools. Our aim was to support the classroom topics by relating them to their application in the world outside school. Biology is applied to food both directly, in the food producing and processing industries, and indirectly, in the research institutes and universities which support those industries. Our bibliography may therefore be more relevant to the teacher than to the pupil as a base for personal literature research. The same may be true of the addresses of the places where some of the extension work is carried out; with courtesy, further information is always forthcoming.

Bibliography

Many of the principles of growing plants and animals do not require up-to-date books; they are established biological principles. What do need to be regularly updated are the economic basis for agricultural practice and the details of present practice and of the advent of new biological knowledge and technologies. To this end, journals and newspaper articles provide ready access to the 'latest' developments.

Of general interest

Journals:
New Scientist.
Scientific American.
World Animal Review. (Food and Agriculture Organisation of the United Nations.)
The Guardian: 'Futures' articles.
Science and Society (1981): (Heinemann with the Association for Science Education.)
Selected titles in the series: *Handbooks in Agriculture* (Longman). It is planned to extend this to some ten or more volumes.
T. Jennings (1974) *Science and Farming*. (Wheaton.)
M.J. Chrispeels, D. Sadava (1977) *Plants, Food and People*. (Freeman.)
All articles in *Scientific American* (Sept. 1976) **235**, no. 3.

Chapter 1

S. Wortman (1976) Food and Agriculture. *Scientific American*, **235**, 30–39.

Chapter 2

C.A. Keele, E. Neil (1971) *Samson Wright's Applied Physiology*. (Oxford University Press.)
M. Pyke (1975) *Success in Nutrition*. (John Murray.)
N.S. Scrimshaw, V.R. Young (1976) The Requirements of Human Nutrition. *Scientific American*, **235**, 50–64.
T.G. Taylor (1978) *The Principles of Human Nutrition. Studies in Biology 94*. (Arnold.)

Chapter 3

W.M.M. Baron (1979) *Organisation in Plants*. 3rd Ed. (Arnold.)
J. Janick, C.H. Noller, C.L. Rhykerd (1976) The Cycles of Plant and Animal Nutrition. *Scientific American*, **235**, 74–86.
E.P. Odum (1975) *Ecology*. (Holt, Rinehart and Winston.)
J.R. Postgate (1978) *Nitrogen Fixation. Studies in Biology 92*. (Arnold.)

Chapter 4

J.C. Bowman (1977) *Animals for Man. Studies in Biology 78*. (Arnold.)
C.D. Darlington (1963) *Chromosome Botany and the Origin of Cultivated Plants*. (George Allen and Unwin.)
C.D. Darlington (1969) *Evolution of Man and Society*, Chapter 4. (George Allen and Unwin.)
J.R. Harlan (1976) The Plants and Animals that Nourish Man. *Scientific American*, **235**, 88–97.
S.G. Harrison, G.B. Masefield, M. Wallis (1969) *The Oxford Book of Food Plants*. (Oxford University Press.)
N.W. Simmonds (1976) *Evolution of Crop Plants*. (Longman.)

Chapter 5

M. Eddowes (1976) *Crop Production in Europe*. (Oxford University Press.)
The Weed Control Handbook. Vol. 1. (Blackwells, Oxford.)
F.L. Milthorpe, J. Moorby (1979) *An Introduction to Crop Physiology*. 2nd Ed. (Cambridge University Press.)
R.D. Park (1974) *Animal Husbandry*. (Oxford University Press.)
E.W. Russell (1973) *Soil Conditions and Plant Growth*. 10th Ed. (Longman.)

J.F. Sutcliffe, D.A. Baker (1974) *Plants and Mineral Salts. Studies in Biology 48.* (Arnold.)
J.S. Willcox, N.W. Townsend (1964) *An Introduction to Agricultural Chemistry*. 3rd Ed. (Arnold.)
C.N. Williams (1975) *The Agronomy of the Major Tropical Crops.* (Oxford University Press.)
Projector slide sets produced by the Glasshouse Crops Research Institute:

No. 1: Nutrient Deficiencies of Cucumber.
No. 2: Nutrient Deficiencies of Chrysanthemums.
No. 3: Nutrient Deficiencies of Carnations.
No. 4: Nutrient Deficiencies of Tomatoes.

Chapter 6

C.R. Austin, R.V. Short (eds.) the series: *Reproduction in Mammals*. (Cambridge University Press.)

1 Germ Cells and Fertilisation
2 Embryonic and Foetal Development
3 Hormones in Reproduction
4 Reproductive Patterns
5 Artificial Control of Reproduction

R.H.F. Hunter (1980) *Physiology and Technology of Reproduction in Female Domestic Animals*. (Academic Press.)
P.F. Waring, I.D.J. Phillips (1970) *The Control of Differentiation in Plants*. (Pergamon.)

Chapter 7

J.C. Bowman (1974) *An Introduction to Animal Breeding. Studies in Biology 46*. (Arnold.)
D. Briggs, S.M. Walters (1969) *Plant Variation and Evolution.* (World University Library.)
D.A. Jones, D.A. Wilkins (1971) *Variation and Adaptation in Plant Species.* (Heinemann.)
K.R. Lewis, B. John (1972) *The Matter of Mendelian Heredity*. 2nd Ed. (Longman.)

Chapter 8

R.W. Allard (1964) *Principles of Plant Breeding*. (Wiley.)
J.C. Bowman (1974) *An Introduction to Animal Breeding. Studies in Biology 46.* (Arnold.)
D.C. Dalton (1980) *An Introduction to Practical Animal Breeding.* (Granada Publishing, St Albans.)

W.J.C. Lawrence (1968) *Plant Breeding. Studies in Biology 12.* (Arnold.)
N.W. Simmonds (1979) *The Principles of Crop Improvement.* (Longman.)

Chapter 9

C.M. Duffus, J.C. Slaughter (1980) *Seeds and their Uses.* (Wiley.)
M. Eddowes (1976) *Crop Production in Europe.* (Oxford University Press.)
D.D. Harpstead (1971) High-Lysine Corn. *Scientific American*, **225**, 34–42.
J. Krummel, W. Dritschilo (1977) Resource Cost of Animal Protein Production. *World Animal Review*, **21**, 6–10.
D. Pimentel, M. Pimentel (1979) *Food, Energy and Society.* (Arnold.)
C.N. Williams (1975) *The Agronomy of the Major Tropical Crops.* (Oxford University Press.)
M. Winstanley (1979) A Cure for Stress. *New Scientist*, **84**, 594–6.
A.W. Speedy (1980) *Sheep Production.* (Longman.)

Chapter 10

Why Feed Pests? A leaflet prepared by the British Agrochemicals Association.
J.M. Cherrett, *et al.* (1971) *The Control of Injurious Animals.* (The English University Press.)
H.F. van Emden (1974) *Pest Control and its Ecology. Studies in Biology 50.* (Arnold.)
B.J. Deverall (1969) *Fungal Parasitism. Studies in Biology 17.* (Arnold.)
Glasshouse Crops Research Institute. Growers Bulletins.
No. 3: Integrated Control of Tomato Pests.
No. 1: Biological Control of Cucumber Pests.
D.L. Gunn, J.G.R. Stevens (eds.) (1976) *Pesticides and Human Welfare.* (Oxford University Press.)
J. Mortelmans, P. Kageruka (1976) Trypanotolerant Cattle Breeds in Zaire. *World Animal Review*, **19**, 14–7.
M.J. Samways (1981) *Biological Control of Pests and Weeds. Studies in Biology 132.* (Arnold.)
M.J. Way (1977) Integrated Pest Control with special reference to Orchard Problems. The Amos Memorial Lecture in the *Annual Report of East Malling*, 1977.
B.E.J. Wheeler (1976) *Diseases in Crops. Studies in Biology 64.* (Arnold.)

Projector slide sets produced by the Glasshouse Crops Research Institute:

No. 5: Pests of Cucumber
No. 6: Pests of Chrysanthemums
No. 7: Pests of Tomatoes
No. 8: Predators and Predation in Glasshouses

Chapter 11

Biotechnology and Education – The Report of a Working Group (1981). (The Royal Society.)
Chemistry in Industry (April 1981) contained a series of articles (pp. 204–49) which coincided with the Second European Congress in Biotechnology in 1981.
W.C. Noble, J. Naidoo (1979) *Microorganisms and Man. Studies in Biology 111.* (Arnold.)
C. Ratledge (1979) Resources Conservation by Novel Biological Processes. 1. Grow Fats from Waste. *Chemical Society Reviews*, **8**, 283–96.
Scientific American, September 1981: devoted to industrial microbiology.
J.E. Smith (1981) *Biotechnology. Studies in Biology 136.* (Arnold.)
Tate and Lyle, *Biotechnology in the 80's.*
I.J. Taylor, P.J. Senior (1978) Single-Cell Proteins: a New Source of Animal Feeds. *Endeavour New Series*, **2**, 31–4.

Chapter 12

Commonwealth Agricultural Bureaux Golden Jubilee Book (1980) *Perspectives in World Agriculture.*
Glasshouse Crops Research Institute Growers Bulletin: *No. 5: Nutrient Film Technique.*
E. González-Jiménez (1977) The Capybara. *World Animal Review*, **21**, 21–30.
A. Huxley (1974) *Plant and Planet.* (Penguin.)
L. Innes (1980) A Baked Bean for Britain. *New Scientist*, **87**, 100–2.
N.W. Pirie (1967) Orthodox and Unorthodox Methods of Meeting World Food Needs. *Scientific American*, **216**, 27–35.

Institutes for agricultural research in Great Britain

Most Institutes have either Information Officers, who can deal with specific or general queries, or open days; some have both.

Many produce summary or information sheets about the work carried out by their scientists which relates to the food-producing and food-processing industries.

Most of these Institutes publish annual reports giving up-to-date review articles and summaries of detailed papers. The value of such specialist use must be balanced against the relatively high cost of these publications.

ARC Institutes

Animal Breeding Research Organisation	West Mains Road, Edinburgh, EH9 3JQ
Food Research Institute	Colney Lane, Norwich, NR4 7UA
Institute of Animal Physiology	Babraham, Cambridge, CB2 4AT
Institute for Research on Animal Diseases	Compton, Newbury, Berks., RG16 0NN
Letcombe Laboratory	Letcombe Regis, Wantage, Oxfordshire, OX12 9JT
Meat Research Institute	Langford, Bristol, BS18 7DY
Poultry Research Centre	Roslin, Midlothian EH25 9PS
Weed Research Organisation	Begbroke Hill, Yarnton, Oxford, OX5 1PF

State-aided Institutes in England and Wales

Animal Virus Research Institute	Pirbright, Woking, Surrey, GU24 0NF
East Malling Research Station	East Malling, Maidstone, Kent, ME19 6BJ
Glasshouse Crops Research Institute	Worthing Road, Rustington, Littlehampton, Sussex, BN 16 3PU

Grassland Research Institute	Hurley, Maidenhead, Berks., SL6 5LR
Houghton Poultry Research Station	Houghton, Huntingdon, PE17 2DA
John Innes Institute	Colney Lane, Norwich, NR4 7UH
Long Ashton Research Station	Long Ashton, Bristol BS18 9AF
National Institute of Agricultural Engineering	Wrest Park, Silsoe, Bedford, MK45 4HS
National Institute for Research in Dairying	Shinfield, Reading, RG2 9AT
National Vegetable Research Station	Wellesbourne, Warwick, CV35 9EF
Plant Breeding Institute	Maris Lane, Trumpington, Cambridge, CB2 2LQ
Rothamsted Experimental Station	Harpenden, Herts., AL5 2JQ
Welsh Plant Breeding Station	Plas Gogerddan, Aberystwyth, Dyfed, SY23 3EB
Wye College, Department of Hop Research	Ashford, Kent, TN25 5AH

State-aided Institutes in Scotland

Animal Diseases Research Association	Moredun Institute, 408 Gilmerton Road, Edinburgh, EH17 7JH
Hannah Research Institute	Ayr, KA6 5HL
Hill Farming Research Organisation	Bush Estate, Penicuik, Midlothian, EH26 0PH
Macaulay Institute for Soil Research	Craigiebuckler, Aberdeen, AB9 2QJ
Rowett Research Institute	Bucksburn, Aberdeen, AB2 9SB
Scottish Crop Research Institute	Invergowrie, Dundee, DD2 5DA
Scottish Institute of Agricultural Engineering	Bush Estate, Penicuik, Midlothian, EH26 0PH

Other sources of information

Agricultural Development and Advisory Service (ADAS) has regional and local offices which are willing to clarify technical points and which may provide pamphlet-type literature at low cost.

Local agriculturally based companies, such as seedsmen and chemical suppliers, are often helpful in providing posters and data about technical products.

Local food-processing companies – such as dairies, cheese factories and breweries – often have facilities for visits by organised parties.

Many of the national food-processing industries and companies have information services. Here are four such:

The National Dairy Council,
National Dairy Centre,
John Princes Street,
London W1M OAP

The Brewer's Society,
42 Portman Square,
London W1H OBB

The Egg Authority,
Union House,
Eridge Road,
Tunbridge Wells,
Kent KN4 8HF

British Agrochemicals Association,
Alembic House,
93 Albert Embankment,
London SE1 7TU

Index